Bicycle Biomechanics

Anthony Crimin • Anthony McGarry

Bicycle Biomechanics

How to Build a Power Dynamometer

Springer

Anthony Crimin
Glasgow, UK

Anthony McGarry
Department of Biomedical Engineering
University of Strathclyde
Glasgow, UK

ISBN 978-3-031-85711-9 ISBN 978-3-031-85712-6 (eBook)
https://doi.org/10.1007/978-3-031-85712-6

This Springer imprint is published by the registered company Springer Nature Switzerland AG
The registered company address is: Gewerbestrasse 11, 6330 Cham, Switzerland

To my wife for her unwavering support and my children for their source of inspirational questioning.

Preface

A fascinating aspect of bicycle design is that it's a design greater than the sum of a single inventor's contribution, like a book with a collective wisdom of many authors. From a technological point of view, the social influence of the bicycle is immeasurable. Socially, the bicycle has given people freedom to travel distances that would not be possible on foot, while at the same time given rise to different genres such as the *Raleigh Chopper* or gravel adventure bikes. Cycling has provided both social and economic influence on our lives. Tubing, wheels, bearings and gears have all influenced the modern world in which we live. From what flies in the sky to the modern motorcar; technology has now turned full circle because technologies developed in other industries that where initially influenced by the bicycle now influence the use of composite materials through to disc brakes now used on modern bicycles.

Indeed, the bicycle has and continues to have a bearing on the direction I have taken in life. Since, taking up cycling around the age of 13, the bicycle has influenced my decision to study engineering and later specialise in biomechanics. In that time, I have seen bicycle technology develop, and along with it, the positions adopted for specialist pursuits such as time trialling, mountain biking, touring and racing.

It was an accident involving a head on collision with a car in my 30s that resulted in me further scrutinising the media "sound bite" publicised in blogs and magazines associated with bicycle technology when finding a path to recovery. The jargon associated with "bicycle feel" and "handling" underpins how the bicycle should be set-up for an individual. This may be influenced by a variety of factors such as fitness; flexibility; riding distance; and type of riding. The geometry has an influence on how the bicycle feels because it has a greater effect on rider biomechanics than on road holding, and this book will endeavour to show the reader that this is the case.

Personally, I like to go as fast as possible for as little possible effort, whether going to work, shops or an epic long-distance ride. The conclusion I have reached is that a bicycle must be set up appropriately for the type of riding undertaken and is influenced by the human factors of physiology and biomechanics.

This publication is not intended to offer a "ready-reckoner" solution for optimal bicycle fit, but instead summaries the story of bicycle fit. From Chaps. 1, 2, 3 and 4 we explore why the bicycle took the form that we see today; how do our lower limbs work with the bicycle; and how does muscle physiology influence our power output. In the final chapter, we investigate how we can measure our riding in a meaningful way. This chapter concentrates primarily on how to build a dynamometer, to measure the kinetics (forces) associated with cycling and may be read as a background to give the reader an understanding of the difficulties of obtaining good measurement in this area. It is hoped therefore that this book will give the reader appropriate background into factors to consider when purchasing a bicycle and where to look if they want to delve deeper into training for speed or simply as a background into human factors of cycling and their measurement.

Glasgow, UK

Anthony Crimin
Anthony McGarry

Acknowledgements

I would like to thank Ian Jones, who is the chair of the *British Society for Strain Measurement* (BSSM) certification committee for his helpful advice in the presentation and content of the final chapter.

I would also like to thank Dr Tony McGarry (the book Co-Author) for his friendship and guided inspiration when studying the field of Biomechanics, whose guided questioning always advanced me to look beyond the data measured outcomes and interrogate the human side of data.

Contacts

The corresponding author contact e-mail is as follows: anthony.crimin@gmail.com

Contents

Acronyms

ADC	Analogue to digital conversion
ADP	Adenosine diphosphate
AI	Artificial Intelligence
ATP	Adenosine triphosphate
BDC	Bottom Dead Centre
CFD	Computation fluid dynamics
COM	Centre of mass
DOF	Degrees of Freedom
EMG	Electromyography
FBD	Free body diagram
HPV	Human Power Vehicles
IDE	Integrated development environment
ISB	International Society of Biomechanics
JST	Japan Solderless Terminal
mph	Miles per hour
rpm	Revolutions per minute
TDC	Top Dead Centre
UCI	Union Cycliste Internationale

Chapter 1
Bicycle Development and Technology

1.1 Introduction

The story of bicycle is over two centuries old and begins with Karl von Drais' velocipede "running machine". To gain an appreciation of why the bicycle was born and its design progression, the priorities of energy efficiency, engineering design and biomechanics should be followed. The form of bicycle that we use today was devised after 100 years of development, and its form has remained to this present day (even though the geometry of the latest frames may not follow the standard diamond structure, the human interface remains the same). The saddle, pedals and handle bars continue to remain in the same place, while changes to the wheelbase, fork rake and trail have allowed bicycles to be more adapted to specific activities.

Recognisable developments of the bicycle are illustrated in Fig. 1.1. The velocipede is like a modern-day toddler balance bicycle, whereby it is propelled by running the feet on the ground.

I would certainly question manufacturers' claims that subtle geometric changes of head or seat tube angle, a degree this way or that, would make a real difference to how the bicycle feels. Required geometric changes are needed for practical reasons, such as how much toe clearance is required, is the bicycle very small or very big, is luggage being carried or is an aerodynamic position a priority. If these practical considerations and design rules of thumb are not adhered to, the bicycle will not feel comfortable. Further advances in aerodynamics, biomechanics and material technologies have resulted in a wide variety of branded bicycles being available, giving more scope to suit rider's anthropometrics, and new lighter materials would easily allow a bicycle to be lighter than the Union Cycliste Internationale (UCI) weight limit of 6.8 kg. These practical aspects of design have allowed bicycle frames to "feel" and "handle" better than the 20 kg safety Rover bicycle—an 1890s modern bicycle.

Hence, the aim of this chapter is to encourage the reader to look at bicycle development with respect to the aforementioned parameters: by gaining a coherent

A. Crimin, A. McGarry, *Bicycle Biomechanics*,
https://doi.org/10.1007/978-3-031-85712-6_1

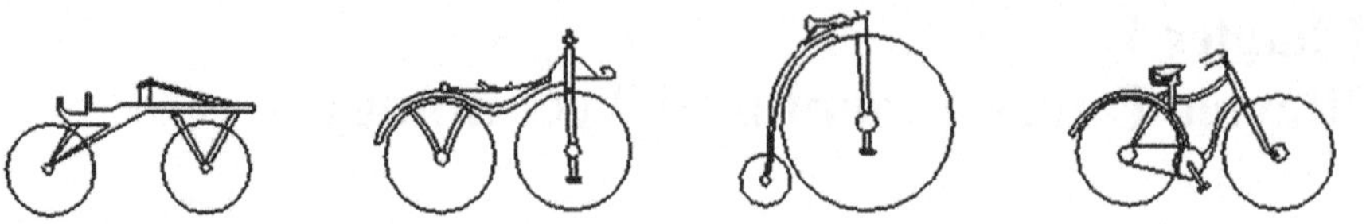

Fig. 1.1 Recognisable developments of the bicycle

understanding of where we are (with respect to bicycle design), a deeper understanding of the biomechanics of cycling can be gained. It is worth looking back to what was arrived at primarily by educated trial and error. The position of our body centre of mass (COM), leg position and trunk orientation with respect to the pedals influence the leverage that out body can use to develop power to turn the pedals, which will be discussed in more detail in Chap. 3.

Furthermore, it is not only our position that has changed on the bicycle, there have also been a variety of mechanisms developed that have been used to transfer forces generated by our muscles to the rear wheel driving force. It will be shown in the subsequent chapters that while the reciprocal motion of modern gearing to an extent allows the muscles to achieve their activation relaxation mechanics for repetitive motion, the gearing of bicycles only allows a small portion (5–20%) of the pedal force to be translated into a rear wheel driving force. Indeed, there have been a variety of lever mechanisms that have both employed the lower and upper body to drive us forward. Many of the early crank designs were directly attached to the rotating wheel which certainly has advantages with respect to the portion of pedal force used to drive the bicycle. Thus, this chapter is not intended to provide the reader with a definitive bicycle history; instead, it follows main design trends from the velocipede to high wheeler, raises questions with respect to the positions adopted (cranks attached to the front wheel), and explores the motivations with respect to some main subsequent design changes. The chapter finalises the discussion by investigating the geometric layout such as rake and trail of the modern-day bicycle.

1.2 To the High Wheeler (1880s)

1.2.1 The Velocipede

In 1818, Drais submitted a patent in Paris where he referred to his running machine as the velocipede after riding 300 miles from Nancy to Paris (Sharp 1896; Hadland et al. 2014). The Drais velocipede was widely copied and was soon imported to England where Denis Johnston acquired a patent in 1818. Evidence suggests that over 300 velocipedes were produced at approximately £10 each (£706 considering

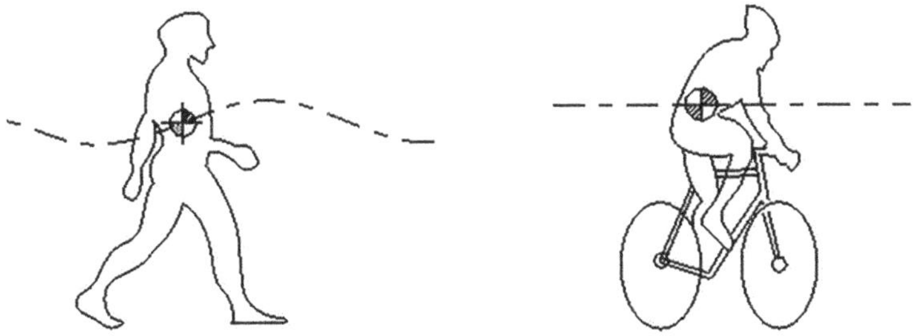

Fig. 1.2 Motion of the body COM during level walking and cycling

inflation at the time of writing) and differed from the original design, in that the steering had no trail or rear brake (Hadland et al. 2014, p. 147).

An interesting question to ponder is what inspired Karl von Drais to balance his body upon two wheels. Why would he think that two wheels would be more (metabolically) efficient than walking, considering that the velocipede weighed up to 20 kg—more than double the weight of a modern-day steel bicycle. Inspiration may have been drawn from observing loads that were more easily moved and hence kindled the idea of self-propulsion.

It is likely that the bicycle was developed from wheelbarrow designs imported from China to Europe in the late eighteenth century (Hadland et al. 2014, p. 3). It is often said that necessity is the mother of invention, and poor harvests of 1816 and 1817, which were followed by those of 1812 and 1813, made horse transportation very expensive at the time. The lifetime cost of owning a horse was at that time approximately £1700, equivalent to over £157,000 today, using the Bank of England inflation calculator (Hadland et al. 2014, pp. 7–9).

Another explanation suggested by Burrows and Hadland (2008, p. 13) is that labourers racing wheelbarrows may have asked the question if two wheels would be more fun. Whatever the rationale, they were brilliantly intuitive from a biomechanical point of view. When observing the body COM when walking, it progresses sinusoidally in a sea of gravity gaining potential and then losing it as kinetic energy. The lower limbs lift the body COM and redirect the fall using the feet so that the body is rolled forward in the required direction of travel. Conversely, despite the additional mass of the bicycle, cycling does not require this exchange of potential and kinetic energy, indicating that lifting the body while walking is the main metabolic cost of ambulation and the primary saving when cycling (Fig. 1.2).

The design progression of bicycle can be viewed by assessing the metabolic cost of evolution as shown in Fig. 1.3. Even though there are subtleties to this narrative, it is easier to go faster and further with reduced metabolic cost for a given speed with each subsequent bicycle design. Compared to walking (Fig. 1.2), the velocipede gave the most significant jump of energy saving with respect to walking (Minetti et al. 2001). However, the important summary to take from the velocipede is that it was more than a running machine; it allowed its users to travel at speeds of 8–10 mph depending on the condition of the road surface (Sharp 1896, p. 147).

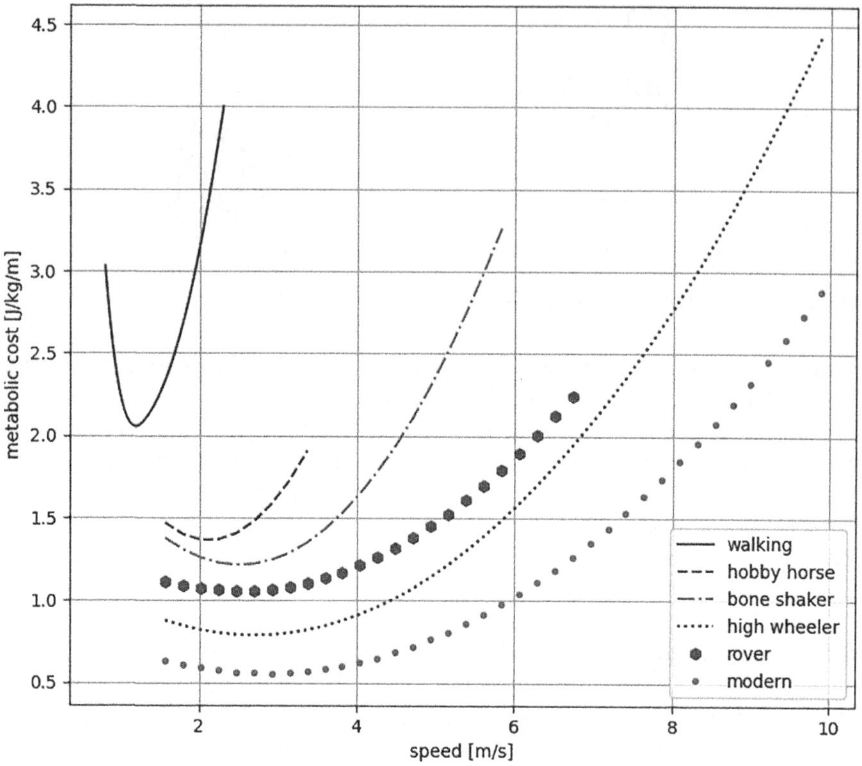

Fig. 1.3 Metabolic cost of bicycle development, plot coefficients from Minetti et al. (2001)

1.2.2 Tricycles, Quadracycles and Roller-Skates

The velocipede found itself out of favour after 1817 as a consequence of the improving economic conditions. Horsepower once again became affordable, due to the first good harvest after several years (Hadland et al. 2014, p. 18). At a cost of up to £10, the velocipede was unaffordable for the working class and became the image of the high-class *dashing dandies.* Collisions (on pavements) did not help. Hence, the perception of being a public nuisance led to the 1835 pavement act (in the United Kingdom, though it happened in other countries), forcing velocipede users onto the road (Hadland et al. 2014). The poor state of road infrastructure, which was not helped by the railways boom of the 1840s, did not improve popularity. It took until the 1870s before road conditions improved, which led to the period where development was mainly concerned with tricycles and quadracycles (Hadland et al. 2008, p. 31).

This development of tricycles and quadracycles primarily came about because of their perceived stability and also gave rise to mechanisms where the riders used their arms, legs or both to propel their vehicle. The Rantoone (Fig. 1.4) won races against

Fig. 1.4 The Rantoone. (Hadland et al. 2014, p. 33)

newly made French velocipedes in 1869 at the crystal palace (Hadland et al. 2014, p. 32). At the time, there were many debates as to whether arm or leg actuator mechanisms were the most effective methods of propulsion. Certainly, there have been many arguments that suggest it is better concentrating on using one set of limbs as the heart only has the ability to provide so much power via the oxygenation of muscles (Burrows and Hadland 2008, p. 19). However, the act of the torso leaning over and pulling on the handlebars or standing on the pedals provides more power for maximal effort. This happens because power is transferred by muscles crossing the body, connecting the upper body (torso) and lower limbs. Therefore, it is difficult to argue that using the lower or upper limbs or both is "better" because it depends on how the limbs can be used; for example, how nimble does the steering need to be, the maximal and average power requirement, duration of effort and the individual's anthropometrics and fitness.

So, why do skates on wheels deserve to be mentioned? It appears that the velocipede may have been the inspiration for other self-propelled personal forms of transport and in turn gave rise to roller-skates. The first roller-skate patents were seen as early as 1819, and this activity continued to grow in popularity throughout the 1850s and 1860s. By 1865, there were roller-skating rinks at Crystal Palace and also evidence that roller-skates were used on the pavements and therefore not considered dangerous like the velocipede (Hadland et al. 2014, pp. 28, 34–36).

Hence, roller-skates gave people confidence to balance their body under wheels, and thus, when it came to the cranked velocipede, they felt assured to lift their legs completely off the ground and use their lower limbs to turn the cranks.

1.2.3 Cranked Velocipede to High Wheeler

The next notable step in bicycle development was the 1860s boneshaker with the cranks directly attached to the front wheel (Fig. 1.5). It is possible to find videos and images on the Internet of people riding boneshakers, and it is clear to see the difficulty of steering and powering the wheel at the same time.

The evolution of the cranked velocipede into the ordinary bicycle and subsequent design variants were possibly a requirement of improving steering (Fig. 1.6). It can be observed that the saddle is positioned more directly above the pedals compared to the boneshaker. Additionally, the larger front wheel of the ordinary bicycle provided greater comfort by being able to roll over obstacles with less vertical translation experienced by the rider (Sharp 1896, p. 243).

Because the cranks are directly attached to the front wheel of the ordinary bicycle, the diameter of the wheel is the size of a gear; one revolution of the cranks results in a single revolution of the wheel. Gearing allows us to change the ratio of a crank revolution to the number of wheel revolutions. Thus a 29-in. wheel will feel like a

Fig. 1.5 French bicycle or boneshaker. (Sharp 1896, p. 148)

Fig. 1.6 Ordinary, high wheeler, or penny-farthing. (Sharp 1896, p. 150)

70-in. gear wheel, if the wheel is geared to make 2.5 turns for a single crank revolution. If you were to build a fixed wheel bicycle, it is likely that by feel, you would select a 70-in. gear, and it is the ratio that you will most likely select on your Derailleur bicycle. Therefore, it is interesting to note that in bicycle shops and club houses across the world, when discussing the gear inch, the effective measure of wheel diameter directly driven is referred to. Moreover, it demonstrates that the average gear that riders used in the 1880s is very similar to the one used now.

However, the position of the body COM with respect to the front hub centre was likely to result in the infamous "headers"—if the front wheel comes to an abrupt stop, the rider would be swung around the large diameter wheel with head first into the ground. As a consequence, variations on a theme of the ordinary bicycle were developed. The advent of the first "safety" bicycles included the Xtraordinary, Facile and Kangaroo (Clayton 2016). All models may be considered the first safety bicycles because their front wheel was reduced to approximately 40 inches in diameter and they used a variety of mechanisms that could provide additional gearing.

Both Xtraordinary and Facile used lever mechanisms to turn the wheel, while the Kangaroo used an offset circular crank system with a separate chain driven by each crank. In addition to being able to select gearing, rider effort was maximised by the reduced frontal area. The effect of the smaller front wheel reduces the height and therefore the drag of the rider. It is recorded in Sharp (1896, p. 153) that

Mr. G. Smith completed a road race on the 26th of September 1884 of 100 miles in 7 h, 7 min and 11 s; by today's amateur standards, it appears particularly impressive.

Conversely, for all that was great about the high wheeler and how they solved problems of gearing and comfort with respect to what came before, from an evolutionally point of view, they were a dead end as only so much could be achieved. Hence, a new form of bicycle began to emerge whose basic layout has not changed for over 100 years.

1.2.4 Invention of the First Pedal-Driven Bicycle

Before briefly discussing the safety bicycle that would ultimately take the form of the current bicycle, we must mention the controversial "World's first pedal bicycle". Strictly speaking, however, this depends on how we define "pedalling". This was not pedalling in the sense we do today—and it may be better to refer to it as the first rear wheel–driven bicycle. At the Glasgow's 1888 International Exhibition, Gavin Daizell was awarded the honour of developing the first pedal bicycle.

However, a relative of the Scottish blacksmith Kirkpatrick MacMillian, James Johnston, campaigned to see MacMillian awarded the honour. It was argued that MacMillian built his bicycle in 1839 or 1840, some 6 years before Daizell and that MacMillian was regularly seen using his bicycle (Sharp 1896, p. 149). However, in truth, Johnston's defence relied upon memories, and given the lack evidence, it is not possible to discount that Daizell was inspired by MacMillian's bicycle. MacMillian utilised existing technologies, and it is documented that the inspiration for the driving mechanism came from William 1788 Symington's steamboat. The rear wheel was turned by crankshafts fixed to the pedals by connecting rods (Fig. 1.7). The clever aspect of this arrangement was that the pedals were not attached to the front wheel, and the rider did not need to sit on top of the driving wheel.

Fig. 1.7 MacMillian's bicycle on show in Glasgow's Riverside Museum

Furthermore, moving the handlebars in front of the rider improved aerodynamics and stability; In the 1890s, Rover safety bicycle was a commercial success, decoupling the steering and driving mechanism.

1.2.5 The Safety Bicycle (1890s)

The rear wheel–driven safety bicycle was invented by Mr. H. J. Lawson in 1879 (Sharp 1896, p. 153). As shown by Fig. 1.8, the rear wheel hub and cranks are on a separate axle, and the handlebars are connected to the forks via a pillar. However, because riders of high wheelers were used to having the handlebars close to them, this design of steering appears to be in part influenced by what came before. It was not long before a new iteration emerged with a sloping front fork (Fig. 1.9) and may have been simply done to move the handles closer to the rider while also giving more toe clearance (Wilson and Papadopoulos 2004, pp. 288–289).

1.3 Bicycle Geometry, Handling, Roadholding and Stability

When flicking through bicycle magazines or websites, the same slogans are seen time and time again such as "confidence inspiring geometry," which gives "high-end road bike performance and all-day ride comfort"; the "sleek aero-profiled design" provides "unlimited speed control regardless of the terrain or road conditions", "a

Fig. 1.8 Rover safety bicycle. (Sharp 1896, p. 154)

Fig. 1.9 Rover safety bicycle with sloping front fork. (Sharp 1896, p. 154)

true one bike to do it all". What does this all mean? Is there really that much difference between frame design?

There is a lot of information on bicycle geometry such as seat and fork angle, wheelbase, frame stiffness, material choice, profile for aerodynamics, and handlebar reach, and these factors may seem to affect how a bicycle feels when we ride it. Certainly, geometric aspects of the bicycle will affect the trunk, thigh and knee flexion/extension angle, and thus, effect muscle power development because of muscle contraction lengths, which will be discussed in greater detail in the following chapter.

As already highlighted in the brief history of bicycle development, the evolutionary designs affect comfort, stability, safety and ability of the rider to transfer limb motion to the bicycle mechanism. All of these designs influence how fast we could ride, and it is clear that wooden wheels with iron rims will result in slower average speed to that of the solid rubber wheels. However, it is interesting to note that it was the velocipede that resulted in the greatest metabolic cost saving relative to walking or running, compared to all the designs which came after. Therefore, with every subsequent design change, the improvements became less pronounced. Consequently, how do the subtle effects of "fit" of the bicycle influence how fast we go? Moreover, the triangular frame that came into being after the first century of design, at the time of writing, is by far the predominant form of bicycle structure. The most radical changes of form compared to the triangular frame are seen with carbon fibre monocoque frames, primarily driven by the need to reduce drag.

The bicycle stays upright with little conscious effort and is one of the reasons why the current design is so enduring. A phenomenal magnitude of research has gone into bicycle balance, and at the time of writing, there is no single equation which relates the balance to the gyroscopic effects of the wheels, the caster effect of the front fork or rear wheel to the steering axis or distribution of the body mass on the bicycle frame. Jim Papadopoulos spent most of his career demonstrating that the rider is more passive in "handling" the bicycle than those manufacturers would have you believe. The tolerance with which the bicycle can balance given geometric differences is much wider than is often publicised and illustrates that there is opportunity for novel designs and is the reason for the current form of bicycle being so enduring (Borrell 2016).

The bicycle is one of those rare transportation devices where the rider is very much heavier than the vehicle, whereas the response of a car or motorbike's handling, roadholding and stability is dependent on its suspension, weight distribution and chassis stiffness. Bicycle riders come in all shapes and sizes with varying ability to manoeuvre. Bicycles don't really handle in a conventional sense for example, they don't have suspension systems that effects their cornering; rather how well they manoeuvre is more influenced by the rider's skill, ability and confidence. Compared to that of a motorbike, its mass distribution, tyres and suspension directly affect the rider's ability to manoeuvre the vehicle as required—this is handling. In comparison, bicycle tyres are so narrow that the rider's ability effects roadholding for a given set of conditions. Obviously, there is a recent trend of road bicycle tyres getting wider, so whilst this is helpful, the stability or ability to maintain an intended manoeuvre is primarily down to the rider's ability (Foale and Willoughby 1988, pp. 35–36).

Why does a new bicycle always feel faster compared to your old one? It is likely that if you are going to spend a considerable sum of money on a bicycle, you will invest time in selecting a frame that is the right size (based of manufacture specifications) and is lighter than its predecessor (given that bicycles are getting lighter and there are more options available than ever before). You are likely to find an off-the-peg bicycle that suits your riding style and physiology. However, despite the considerable number of options available, a bicycle can still be adjusted incorrectly resulting in too much elbow or knee extension (the most notable incorrect adjustment I observe as riders try to obtain the most aerodynamic position possible). However, from the author's point of view, despite geometry affecting the biomechanics of riding and having a more profound effect on the handling or feel of a bicycle, most riders will assume that the feel or handling is affected by the roadholding ability of the bicycle and in turn is influenced by subtle geometric changes. The aspect of how bicycle geometry allows us to "handle" the bicycle and the variations that are commonly found which affect the feedback from the road will be explored. There are clearly rules of thumb that should be followed for toe clearance, characteristics which let you attach a rack and panniers and sit forward to use tribars for time trailing, although all this has little effect on how the bicycle feels.

1.4 Bicycle Geometry and Balance

A riderless bicycle if pushed from a standing start by the saddle will weave forward in a trajectory of self-balance. If there is a puddle on the ground and the road is dry ahead, the weaving motion of self-balance is mapped out, with the rear wheel trailing the front wheel (Fig. 1.10).

This motion was commented on as early as 1869 by the Scottish Engineer William Rankine, and the model devised by Francis Whipple in 1899 is one of the most iconic. The model is composed of the wheels and frame with a point mass and front fork—the wheels are connected by axels to the frame and a hinge joint is used

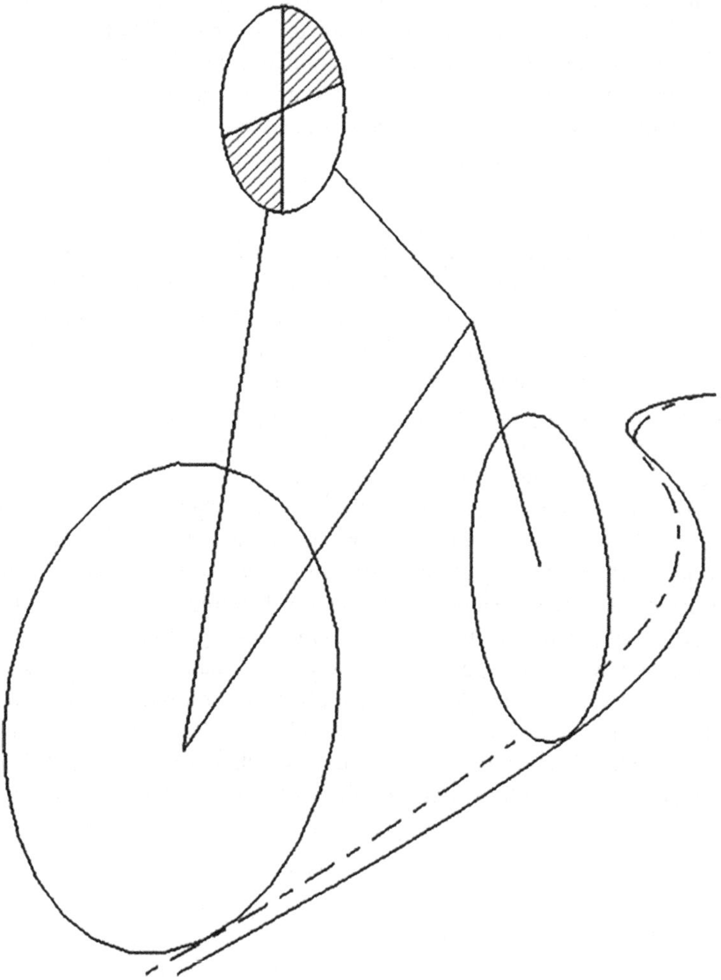

Fig. 1.10 The weave and balance of a bicycle

to connect the frame and fork (Borrell 2016). Fortunately, it is beyond the scope of this book to digest this complex publication, and if you find it online you will see why!

In essence, as demonstrated by observing a riderless bicycle, by briefly steering right as we are turning left, the fall of our COM to the left is more easily controlled because we can steer under the falling COM (to the left) by catching up to the body COM falling to the left by taking the initial right-hand manoeuvre (Hadland et al. 2014). The bicycle is a unique mechanical assembly which requires no control input to keep it upright. Conversely, consider balancing an inverted broom on your hand; if it falls forward away from you, acceleration of your hand under the broom will support its mass and keep it balanced. It is likely you will overshoot, and it will start falling towards you, and by pulling your arm back, it will keep it balanced. Countless engineering systems, whether electrical, mechanical or more likely a combination of both, from air conditioning to self-stabilising ambulating robots, use this method of speed of response, overshoot and damping to control the system.

The bicycle mass distribution, the gyroscopic effects of the wheels and the hinging effect of the front forks combined with the caster effect of the fork and rear wheel to the steering axis allow this mechanical system to be self-regulated—it is inherently stable. Even when the mass of a rider is included, the bicycle can still stay upright moving along a "straight" line without using hands, illustrating the level of tolerance within the system. The equations of motion are not easily distilled down into a design handbook. However, there are certain guidelines that frame builders, bicycle manufactures and even cyclists can follow when, respectively, building or purchasing a bicycle to ensure that the geometry is best suited to their requirements.

1.4.1 Trail

The geometric identities that are most often published by manufactures to tell the buyer how the bicycle will handle are illustrated in (Fig. 1.11). Lengths such as reach, standover and effective height are not shown here as they are related to body anthropometrics.

Because both the front and rear tyre contact the ground behind the steering axis, this gives rise to the positive caster effect. The caster effect gives the bicycle straight line stability and is seen on both motorcycle and car steering geometry. As illustrated in Fig. 1.12, the friction force, respectively, acting on the front and rear tyres F_f and F_r tends to "straighten" the wheels in the direction of motion. The slip angle $\varnothing$ or β can be used to resolve the normal component of friction force as shown in Eqs. 1.1 and 1.2 acting on the tyre contact patch to straighten the wheel. It should be noted that the mechanical trail reduces the height of the bicycle as the steering is turned; this can make the steering feel heavy if the height drops too much.

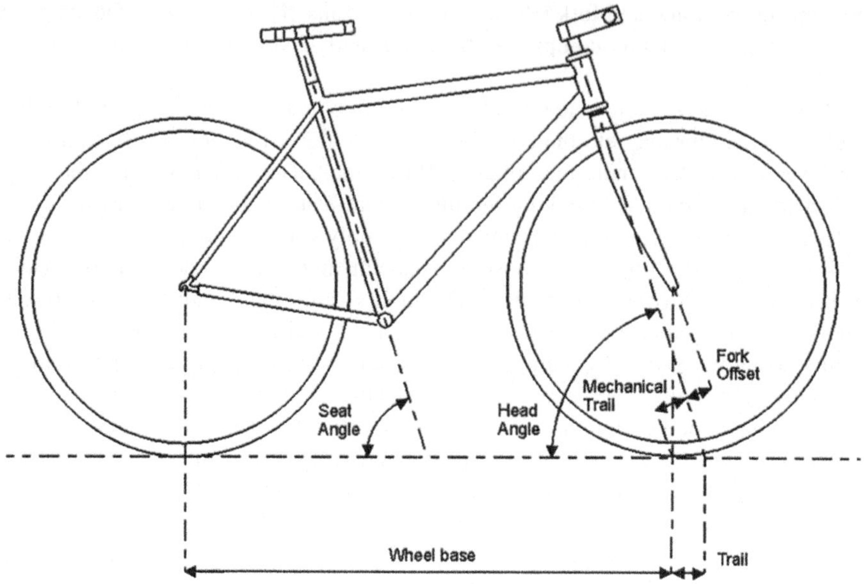

Fig. 1.11 Bicycle handling geometry

$$M_{\text{front}} = lF_f \sin\varnothing \tag{1.1}$$

$$M_{\text{rear}} = LF_r \sin\beta \tag{1.2}$$

Despite the trail of the rear wheel having a longer lever arm, the component of the normal fractional force is a lot smaller by comparison, as a consequence of the much-reduced slip angle. Therefore, the straight-line stability is more greatly affected by fork trial. Thus, it can be intuitively seen why the mechanical trail of touring bicycles (≈55 mm), compared to road racing bicycles (≈36 mm), or even track bicycles (≈23 mm) is greater.

This reaction force explains why the wheel can fall under you when riding on a wet slippery surface, as the forward motion of the bicycle is made up of a series of side-to-side oscillatory curves and friction is required to stop slip. However, it is difficult to objectively quantify the feel with a steering torque value because the friction depends on tyre pressure and width, road conditions, wind loading, wheel construction, weight of rider, position of rider or are there front panniers to name but a few variables. So, what does this all mean? Yes, frame builders or manufactures should follow design guidelines—how does the difference affect the average rider when comparing "handling"—(probably not by very much!) Consider that your bicycle headset has too much preload set or, is even slightly worn, the "stickiness" prevents the bicycle self-balancing: the natural oscillations in the direction of motion

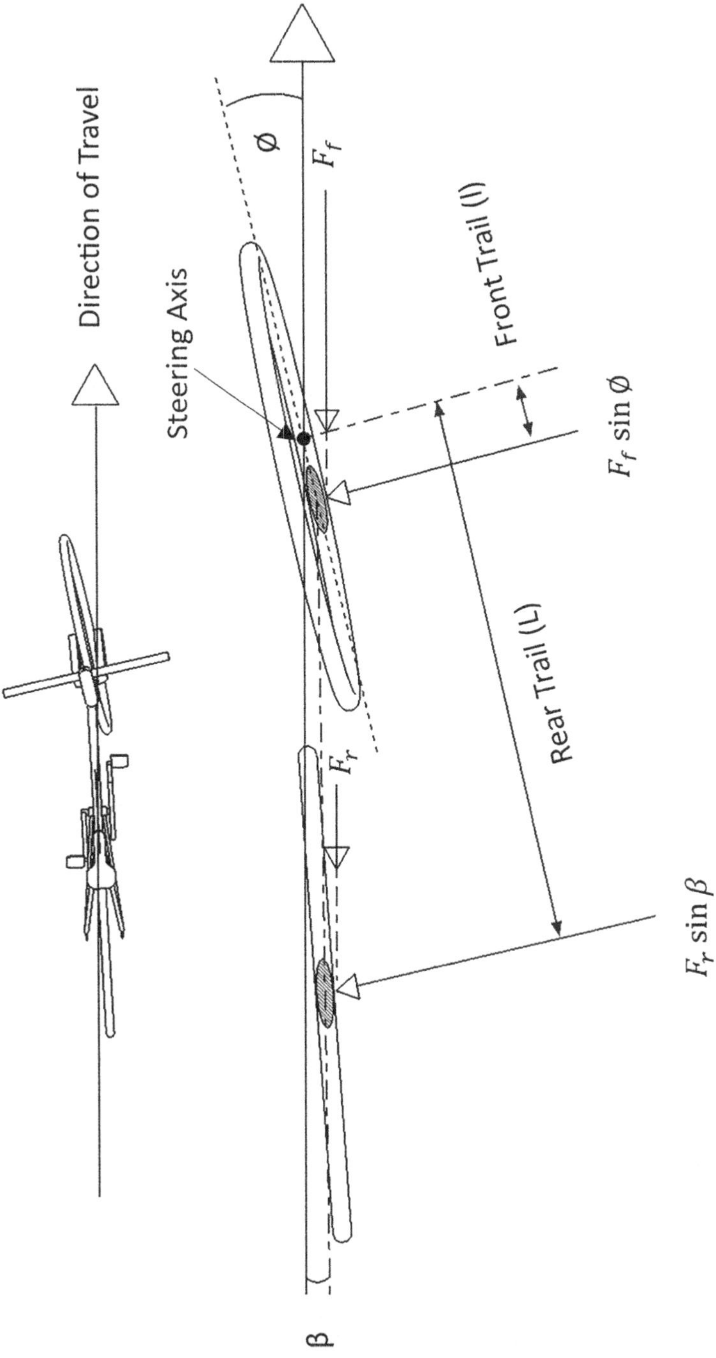

Fig. 1.12 Plan view of bicycle turning. (Inspired by Foale and Willoughby 1988, chap. 2)

are prevented. This gives the reader a feel for the level of friction force that acts to provide balance due to the positive caster effect.

1.4.2 Rake Angle

The difference between Figs. 1.8 and 1.9 (p. 9 & 10) is evident; the angling of the steering axis on the rover design was a clever change which moved the handles closer to the rider and the front wheel away from the riders' feet without the need for linkage.

Usually, racing bicycles have steeper steering angles (74°) compared to touring or audax geometry (72°) as more toe clearance is required for mudguards or panniers. This can affect the wheelbase and the "slacker" head tube angle for non-racing machines and increases the wheelbase, which makes the frame less agile but provides greater straight-line stability.

As illustrated in Fig. 1.13 (p. 17), on increasing the headtube angle to 73° from 72° with no fork offset, the trail and toe clearance is reduced and the tyre is moved closer to the down tube. Clearly, the bicycle will be far safer to ride if offset is incorporated into the forks to move the wheel forward. Conversely, Fig. 1.14 highlights the difference on keeping the head tube angle at 72° and incorporating and fork offset of 45 mm. The toe clearance and frame clearance have been increased, although the rake has been reduced. The dashed line shows the outcome of slacking the headtube angle to 71°—the trail has been increased.

These simple illustrations highlight that the dimensional changes made to bicycle frames are done so for practical reasons. Is the bicycle being used for roading racing, time trials, mountain biking, gravel or touring?—all these adaptations influence the need for ground clearance to how aerodynamically positioned the rider needs to be. In the author's opinion, they do not affect the bicycle handling—although the exception would be full suspension bicycles, they are in their own category. However, for most bicycles, there may be different feedback with respect to how the steering feels; although a nice new lightweight frame that helps you go up the hills faster will indeed make your bicycle feel "better" than before.

1.5 Conclusion

The reader is advised to read books such as Hadland et al. (2014), Clayton (2016) or Ritchie (2009) for a more in-depth synopsis of bicycle history, especially the latter for a thorough cross reference history of MacMillian's bicycle. However, the purpose of this chapter is to show the motivations for the birth of the bicycle—affordable economical transport. Also, while there is something indelible about the "classic" bicycle protected by the UCI, it should not prevent us from looking forward and seeing a whole new class of human power vehicles (HPV).

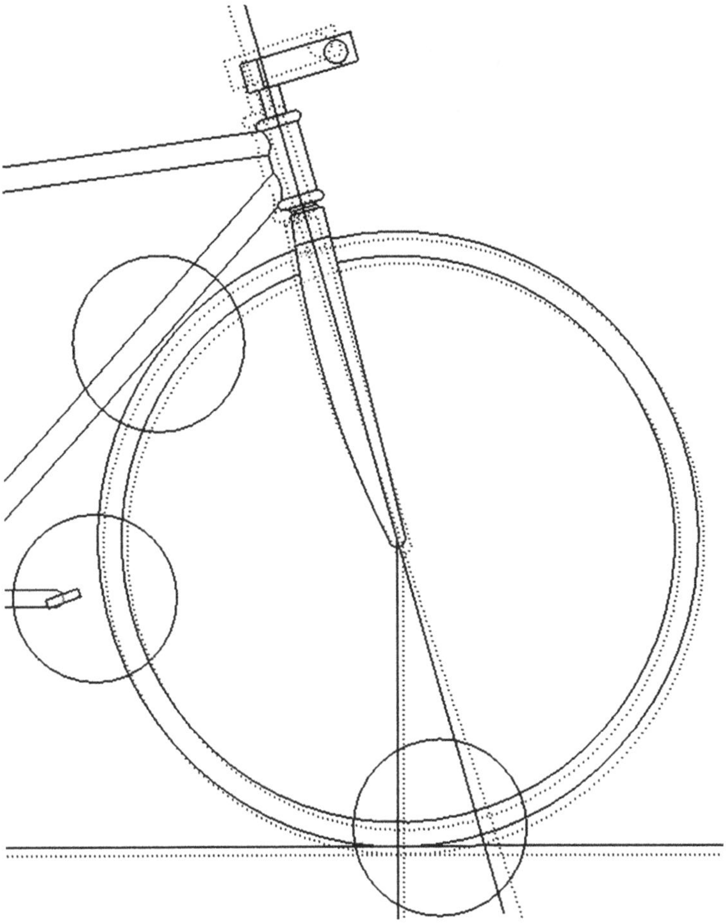

Fig. 1.13 Headtube angle 73° (solid black line) and 72° headtube angle (dotted black line)

Nevertheless, as the bicycle as we know it enters into another century, innovations are driven by adaptation, for example, through aerodynamics, mechanisms, biomechanics and material technologies with new methods of manufacture such as 3D printing. This has resulted in the bicycles becoming more adapted for their intended purpose such as road racing, off-roading or urban commuting. It is interesting to note that in cycling media, these adaptations are seen as new bicycle design, but they are not; however, you cannot disagree with the fact that paddle shifters to disk brakes have certainly improved the riding experience. Ultimately, to improve how your bicycle feels, frame selection is important because fit effects our biomechanics and how the frame feels or handles. The geometry effects practical aspects of frame design and has much less effect on road handling than is intuitively reasoned.

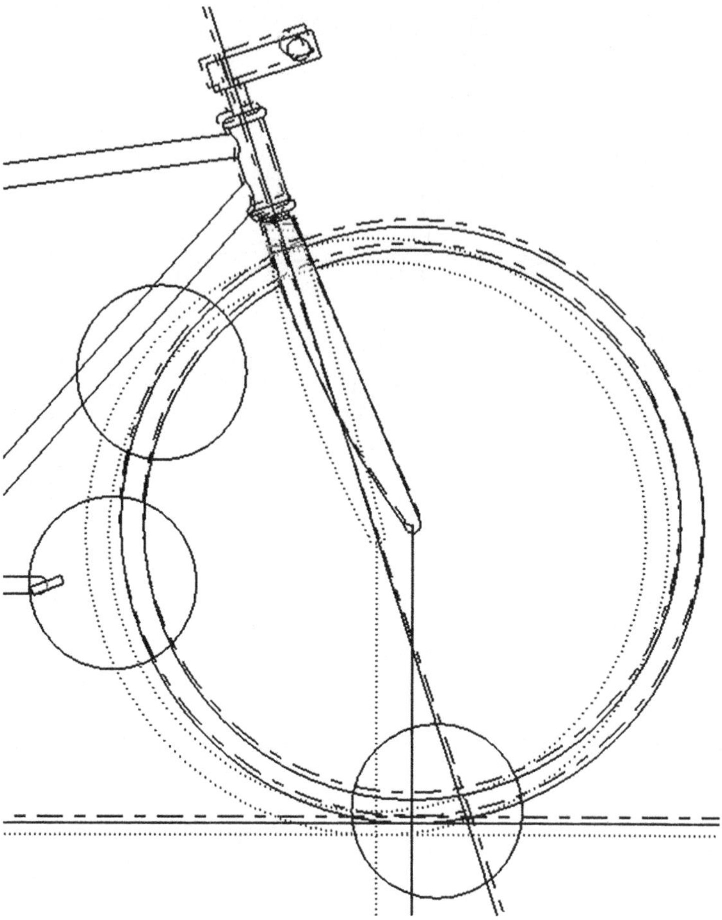

Fig. 1.14 Headtube angle 72° with no offset (dotted line), headtube angle 72° 45 mm offset (solid line), headtube angle 71° 45 mm offset (dashed line)

Bibliography

Borrell B (2016) The bicycle problem that nearly broke mathematics. Nature 535(7612):338–341

Burrows M, Hadland T (2008) Bicycle design: the search for the perfect machine. Snowbooks, London

Clayton N (2016) The birth of the bicycle. Amberley Publishing, Stroud

Foale T, Willoughby V (1988) Motorcycle Chassis Design: The Theory and Practice. Osprey, Cortez

Hadland T, Lessing H-E, Clayton N, Sanderson GW (2014) Bicycle Design: An Illustrated History. MIT Press, Cambridge

Minetti AE, Pinkerton J, Zamparo P (2001) From bipedalism to bicyclism: evolution in energetics and biomechanics of historic bicycles. Proc Biol Sci 268(1474):1351–1360

Ritchie A (2009) The origins of the bicycle: Kirkpatrick MacMillan, Gavin Dalzell, Alexandre Lefebvre: documentation, memory, craft tradition and modern technology. John Pinkerton Memorial Publishing Fund

Sharp, A. (1896). Bicycles & tricycles, an elementary treatise on their design and construction. Lond. &c, The MIT Press

Wilson DG, Papadopoulos J (2004) Bicycling science. MIT Press/IEEE Xplore, Cambridge/ Piscataqay

Chapter 2
Drag and Aerodynamics

2.1 Introduction

At the time of writing, the current land speed record for a human-powered vehicle (HPV) is 89.6 miles per hour (mph), and this is primarily achieved by improved aerodynamics. Essentially, this HPV is a fair recumbent bicycle. However, the speed has come at cost of visibility, balance, mechanical efficiency, manoeuvrability and thermodynamics—so you would not use this vehicle to go to the shops or race in the Tour de France. However, the important note to take from this phenomenal speed is the role that aerodynamics plays on the speed of the bicycle.

In 2002, the Canadian Manufacture Cervélo lunched the *Cervélo S3 road bike*. Initially manufactured from aluminium tubing produced for light aircraft with its aerodynamic teardrop downtube and sweeping features, it was considered a game changer. It was subsequently used in the Beijing 2008 Olympics where it achieved second, third and fourth place in the men's road race. Even though aerodynamic bicycles were used prior to this, these frame types were often banned by the Union Cycliste Internationale (UCI); the Cervélo S3 was different in that it initiated a change that has not lost momentum. At the time of writing in 2023, winner Jonas Vingegaard's average speed of 41.12 km/h (25.55 mph) in the Tour de France represented a trend of increasing tour winning average speeds. Indeed, it is generally believed that the technological incremental improvements year-over-year are making this difference. As manufacturers provide general statements with respect to their designs and differences they make (for example, a profile is more aerodynamic), the wider purchasing public do not realise that the frames they see winning prestigious races are aerodynamically tweaked to fit riders such as Jonas Vingegaard. Moreover, the body suits that are worn by riders have stitching in the right place to "trip the air" to reduce the effect of the low-pressure turbulent wake. Hence, the same bicycle given to you or me is not going to achieve the same aerodynamic performance.

So where does biomechanics and power output fit into the story of aerodynamic performance? Because at least 80% of the aerodynamic drag is a result of the rider,

A. Crimin, A. McGarry, *Bicycle Biomechanics*,
https://doi.org/10.1007/978-3-031-85712-6_2

assuming positions that improve aerodynamic performance should also be balanced with how the musculature develops and transfers power to the pedals. There is a compromise and having an awareness as such is not just beneficial to the elite or amateur racer but for all who want to ride their bicycle further, faster and more comfortably.

2.2 The Propulsion Force

To understand the magnitude of the drag forces that act on us when riding it is first worthwhile exploring the forces that we create to propel us (Burke 1986, chap. 5). When sitting on a stationary bicycle, all the forces act vertically: as illustrated by free body diagram (FBD) in Fig. 2.1. The weight of the rider is supported by the pedals, handlebars and the saddle. The ground reaction force (R_{R} and R_{F}) supports the weight of the rider and bicycle as summarised by Eqs. 2.1 and 2.2.

$$\sum \begin{bmatrix} F_x \\ F_y \end{bmatrix} - \begin{bmatrix} ma_x \\ ma_y \end{bmatrix} = \begin{bmatrix} 0 \\ 0 \end{bmatrix} \tag{2.1}$$

$$\begin{bmatrix} 0 \\ R_R \end{bmatrix} + \begin{bmatrix} 0 \\ R_F \end{bmatrix} = \begin{bmatrix} 0 \\ W_r \end{bmatrix} + \begin{bmatrix} 0 \\ F_h \end{bmatrix} + \begin{bmatrix} 0 \\ F_{pr} \end{bmatrix} + \begin{bmatrix} 0 \\ F_{pl} \end{bmatrix} \tag{2.2}$$

However, when the bicycle is in motion, the external drag forces acting on the bicycle and rider act horizontally against the bicycle. If riding at a constant speed, there is minimal acceleration, so no inertial force will act. However, it will result in subtle adjustments in acceleration to maintain bicycle speed, even if not sprinting. Another force that resists motion includes road-rolling resistance; it can be observed that when the bicycle is in motion, the front wheel rolling resistance acts against the direction of motion, and that the driving force of the rear wheel acts to push the bicycle forward.

$$\begin{bmatrix} R_R \sin \theta R_R \cos \theta \end{bmatrix} + \begin{bmatrix} -R_F \sin \theta \\ R_F \cos \theta \end{bmatrix} = \begin{bmatrix} 0 \\ W_r \end{bmatrix} + \begin{bmatrix} 0 \\ F_h \end{bmatrix} + \begin{bmatrix} 0 \\ F_p \end{bmatrix} + \begin{bmatrix} F_d \\ 0 \end{bmatrix} \tag{2.3}$$

To estimate the horizontal component of driving force that acts on the rear wheel, another FBD is required to investigate the leverage between the cranks and the driving torque of the wheel (Fig. 2.1).

$$F_R \frac{d_w}{2} = F_c \frac{d_g}{2} \tag{2.4}$$

$$F_c \frac{d_c}{2} = F_p r_c \tag{2.5}$$

By rearranging Eqs. 2.4 and 2.5 to equate F_c, the driving wheel reaction force F_{RH} can be described as a function of the pedal force F_p (Burke 1986, chap. 5).

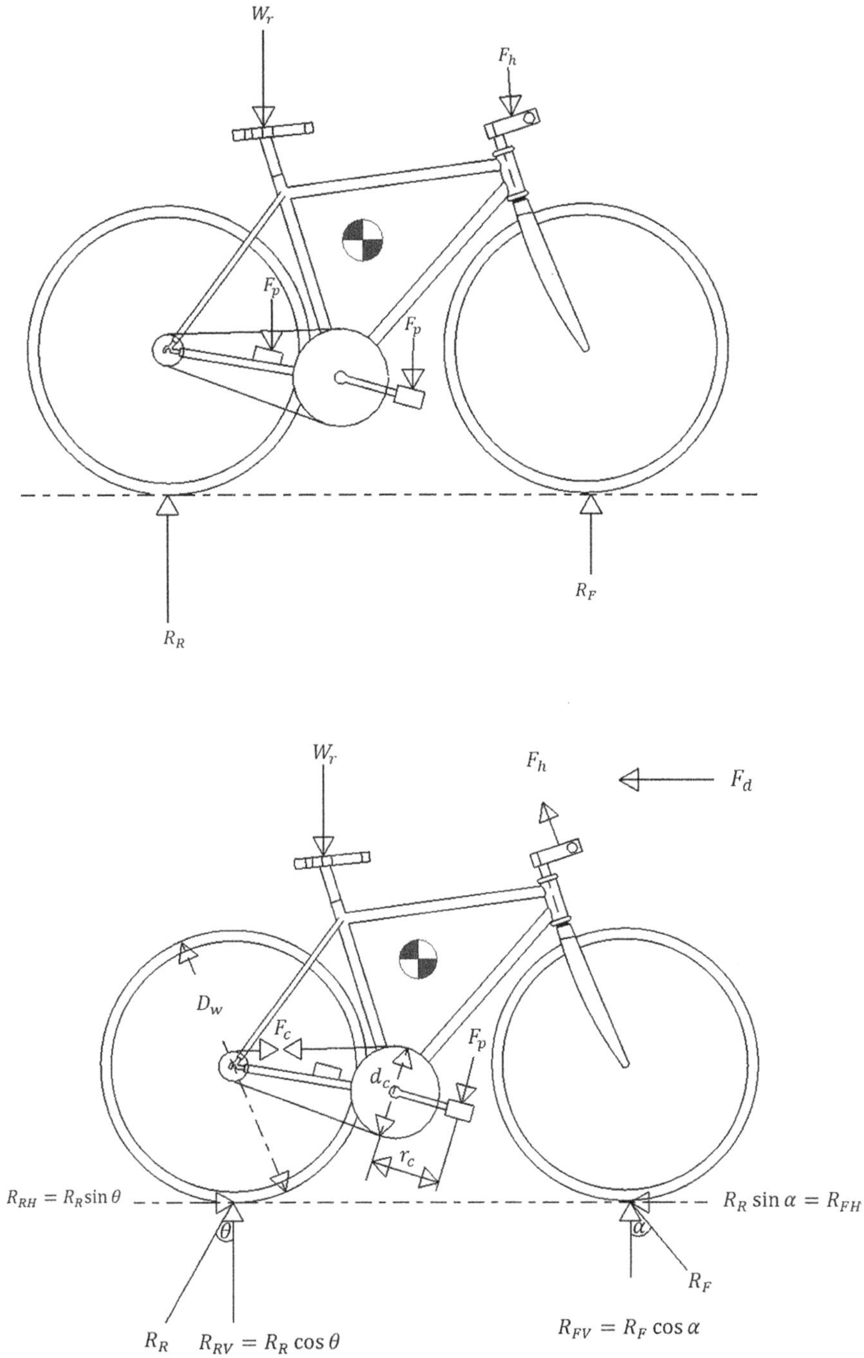

Fig. 2.1 Forces that act on the bicyclist when (top) stationary and when (bottom) riding

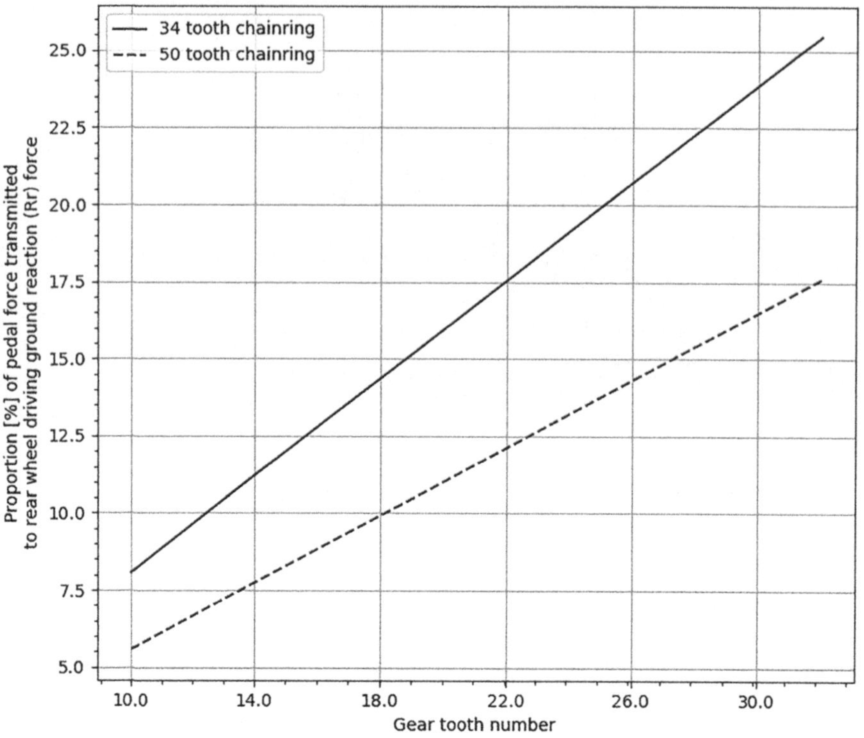

Fig. 2.2 The percentage pedal force that propels the rear wheel

$$F_{RH} = \frac{2d_g r_c}{d_c d_w} F_p \tag{2.6}$$

By plotting the function F_{RH}, it is shown that the gears significantly reduced the leverage effect of the pedal crank system; typically, on flat terrain, F_{RH} will be lower than 10% of the pedal force. As all off-road cyclists are aware, going to a lower gear does not help traction because the wheel torque is increased, and when the ground is soft, this leads to the rear wheel losing traction. However, the main point to consider from this simple analysis is that in reality, only 10–30% of your pedal force (normal to the cranks) is translated into a propulsive force; hence, if your pedal force is 500 N, when using a 50/10 tooth gear combination, your propulsive force is likely to be less than 50 N (≈5 kg).

Thus, this explains why small percentage drag reduction or even a well lubricated chain can make a difference in terms of average speed. It also highlights why changing to a "lower gear" when cycling up a hill helps; this is because more pedal force is used to propel you, and the trade-off is that you go more slowly. Equation 2.7 illustrates this: the minimum and maximum power you can develop is constricted to a fairly narrow range; consequently, if the torque increases, the cadence has to decrease and vice versa (Fig. 2.2).

$$P = T\omega \tag{2.7}$$

2.3 One-Hour Record and its Relevance

When Eddy Merckx broke the 1-h record in 1972 by covering a distance of 49.431 km, the approach his team took was to build the lightest possible bicycle. When Francesco Moser took on the challenge of beating Merckx's record, his team built the most aerodynamic bicycle possible, and he covered a distance of 51.151 km. Merckx's bicycle weighed 5.75 kg compared to Moser's 9.6 kg machine; the drag coefficient of body suit adjustments, the effect of the handlebar height and cranks lengths were a few of the variables that were considered (Monte et al. 1987).

Fast forward to today at the time of writing, and to get an appreciation of the mantra that everything matters when maximising the speed, we only need to observe the latest equipment and sportswear; certainly, in the peloton, the media image is of concern—but manufactures are competing with each other to sell bicycles; is there that much difference? As discussed by Smethurst and Findlay (2023), there is limited evidence that adopting an overly aerodynamic position actually works and has implications for safety with respect to visibility.

However, when using 50–13 tooth gear combination riding at 45 km/h, while the hands are resting on the hoods, Table 2.1, p. 26, a component which allowed, for example, a 3% drag reduction (on a still day) would reduce the drag force by 1.3 N and the power requirement by approximately 19 W (Fig. 2.3).

Granted that this combination of gear and position of sitting on the hoods would not be adopted for cycling at 45 km/h, it illustrates why a small percentage reduction in drag can be a significant saving and also highlights that manufactures' claims that components can reduce drag by more than fractions of a per cent are overexaggerated because such estimated savings of 19 W would give riders more than a significant edge. Indeed, predicting the drag performance of components for indoor track cycling is more reliable than on the road, and considering that 80% of drag results from rider, it is hard for manufactures to predict the performance out with the wind tunnel.

Hence, while drag is import in the right set of conditions/environment, especially when considering an attempt on the 1-h record or a time trial on a still day outside it can be reduced with some confidence as the conditions can be more easily wind tunnel replicated. The varying results of test data between wind tunnel studies, for reasons such as environmental conditions and the uniform flow direction when evaluating components, the percentage reduction in drag claimed by manufactures is not likely to be experienced (Crouch et al. 2017). Frame weight savings will provide a more repeatable and predictable performance gain for everyday cycling. Conversely, when cycling up an incline, for example, of 8% (4.6°), gravity now

Table 2.1 Drag coefficients from Gibertini and Grassi (2008) for projected front areas illustrated

Position	Stem	Brake hoods	Drops	Time trial
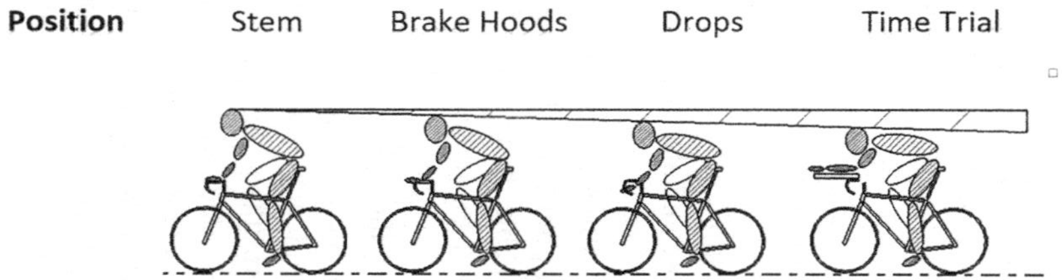				
Projected area (m^2)	0.386	0.400	0.355	0.297
Drag coefficient (C_d)	0.824	0.760	0.814	0.792

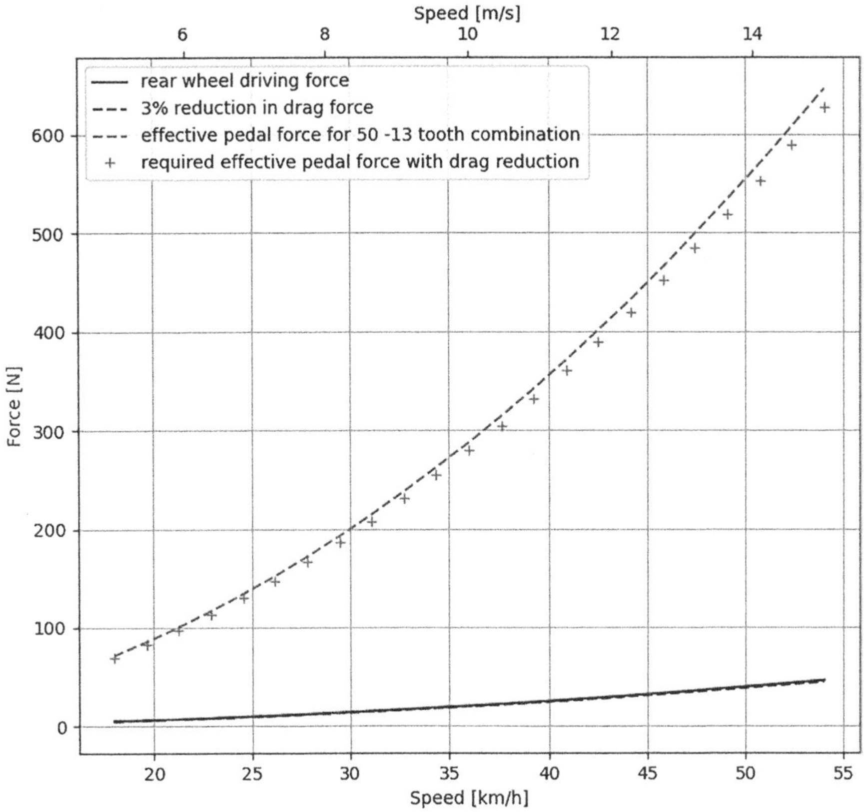

Fig. 2.3 Effect of the 3% reduction on drag has on the effective pedal force and the rear wheel driving force

becomes the main resistance to motion, even to cycle at a speed 20 km/h (12.5mph) is at the limit of sustain human effort on this gradient (Fig. 2.6, p. 30). For example, a 400–500-watt output is required to break the 1-hour distance record (Malizia and Blocken 2020).

A final interesting point to note, to conclude this section, is that our muscles work with impulses and the plots above show average power. Figure 2.4 is the effective pedal force for the author cycling and reveals that despite peaks of 400 N, the average force developed is much less. This highlights the force magnitude that muscles need to develop—and this is not just seen in cycling. Hence, this plot demonstrates the importance of understanding your biomechanics and muscle physiology.

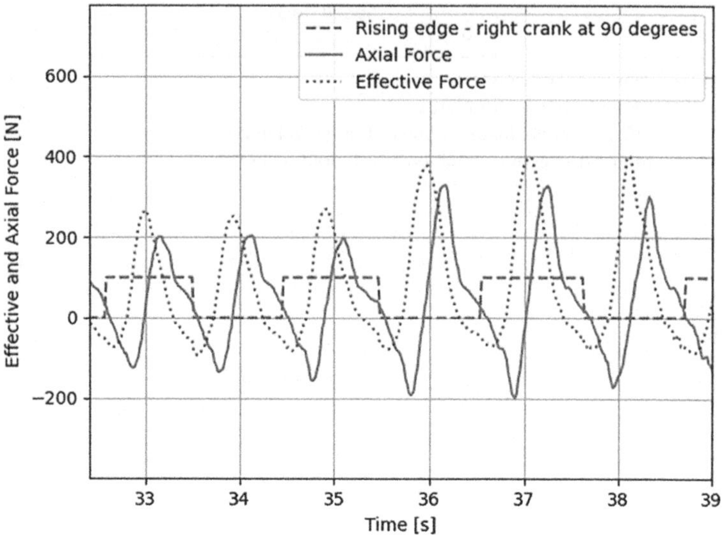

Fig. 2.4 The time-varying pedal input force by the author

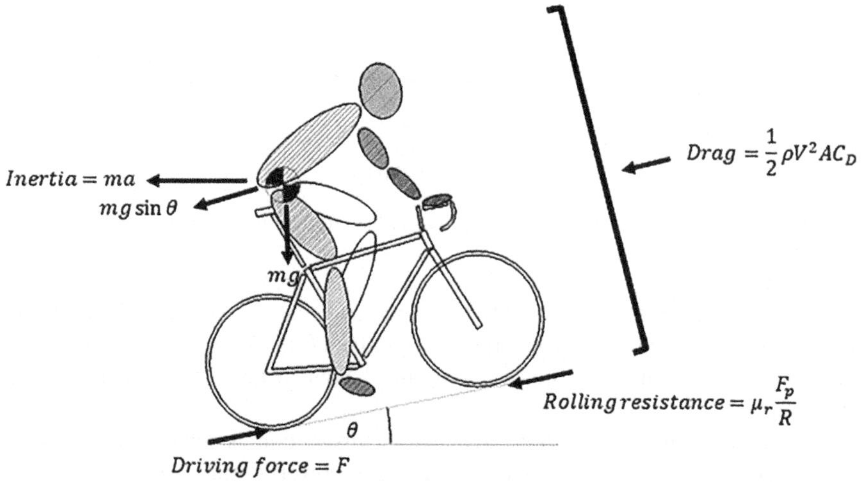

Fig. 2.5 Mechanical factors effecting speed

2.4 Mechanical Factors Effecting Speed

Drag is the only resistance force acting that varies as a function of speed. However, there are other variables that can be considered which resist motion as summarised in Fig. 2.5 and Eq. 2.8.

$$F - \mu_r \frac{F_p}{R} - \frac{1}{2}\rho V^2 A C_D - mg \sin\theta - ma = 0 \tag{2.8}$$

Rolling resistance generally remains constant for a given speed and is affected by how much energy the tyre can return to the bicycle as the tyre deforms and regains its shape. The properties that effect the rolling resistance include air pressure, tread pattern and rubber properties; however, rolling resistance is primarily a function of the weight supported by the tyre. Typically, rolling resistances for bicycle tyres are usually published (https://www.bicyclerollingresistance.com) in watts for a given tyre pressure *and can* be anywhere between 2 and 10 N. If an average 200 N pedal force (Fig. 2.4) and 10 N rolling resistance are considered, these account for 5% of the total resistance to forward motion. This is why the trend of riding with wider tyres on road bicycles has a limit; even though it is more aerodynamic, it increases rolling resistance. However, on a positive note, without friction, we could not have wheel traction and forward motion!

Another consideration of propulsion is the mechanical efficiency of the mechanisms and linkages, which can fall anywhere between 98.6% and 81% (Sneiderman 1999). Therefore, with good chain alignment and maintenance, less than 2% of the transmitted force is lost to mechanical inefficiencies and may explain the trend in recent years to use larger chainrings and cassette diameters so that the chain loses less energy, resulting from turning on smaller gear radii.

The inertial force that a cyclist experiences is a direct result of their mass and acceleration: the harder a cyclist accelerates, the more resistance the rider feels to that acceleration. Furthermore, the limbs will also experience rotational resistance when trying to accelerate. If a rider is moving at a steady speed, they will experience no inertia forces, essentially $ma = 0$ as shown in Eq. 2.8, and the greater the number of accelerations you make, the more resistance you need to overcome. The only practical technique of reducing the inertial forces is to reduce the weight of the bicycle or yourself!

Gravity in hilly terrain is a significant factor, and more so than drag with respect to the resistance to motion. This can be visually represented by using Eq. 2.9 to plot the power output required by the rider to propel themselves for a given speed (Fig. 2.6). The example plots the power required to cycle up an 8% (4.6°) gradient including rolling resistance and drag for a given riding speed on a still day. It clearly emphasises even with a head wind, weight becomes a dominant factor when ascending.

$$P = FV = \left(\mu_r \frac{F_p}{R} + \frac{1}{2}\rho V^2 A C_D + mg \sin\theta\right) V \tag{2.9}$$

Drag or wind resistance is even present on a still day: the faster you go, the more resistance you will feel; unfortunately, the power requirement increases by the velocity cubed (Eq. 2.9). In many ways, it is the most difficult resistance to predict and characterise, and the next section of this book provides a visual and descriptive explanation of why air causes so much resistance.

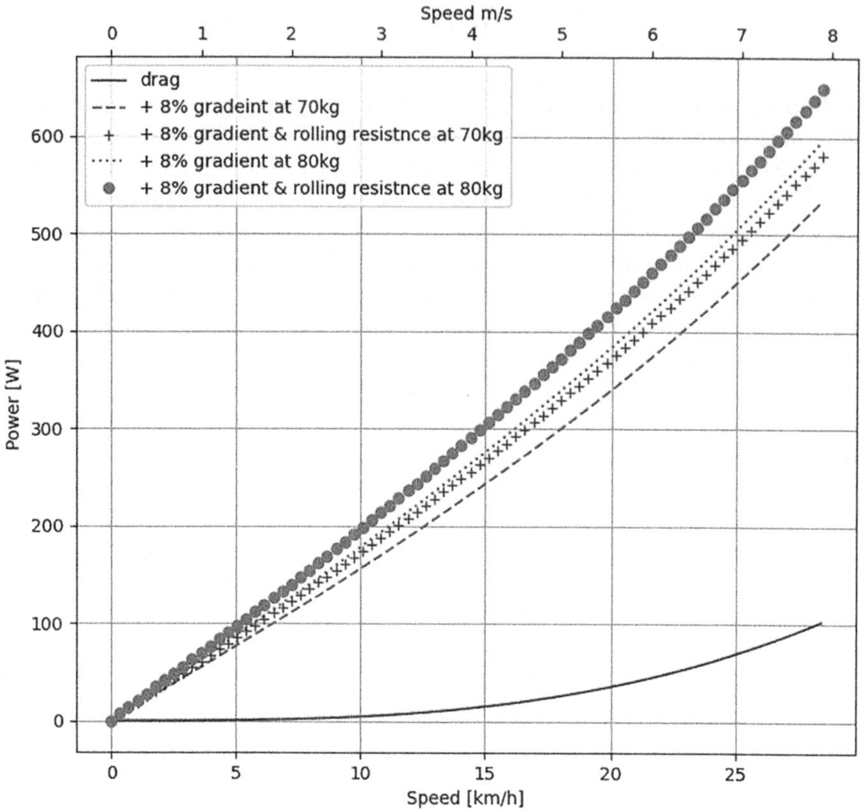

Fig. 2.6 Graphical illustration of rider power output with respect to drag, gravity and rolling resistance

2.5 Aerodynamics

2.5.1 Definitions and Sources of Drag

Unlike road vehicle aerodynamic design, the driver is contained with the body of the vehicle, and this allows for more repeatable results when analysing drag. Since the time of Moser's 1-h record, it has been clearly documented that wind tunnel testing was, and still is, used to best fit professional riders to their machines. The rider accounts for the greatest proportion of the drag force when considering that only 20% of aerodynamic resistance is accounted for by the bicycle (Monte 1987). However, it is worthwhile exploring the type of drag that the bicycle rider experiences and understanding some analytical expressions as well as the difficultly of using results derived from wind tunnel analysis.

When considering drag and lift, there are analytical expressions and dimensionless terms that can be used to relate the shape and speed of the object to a force which the rider needs to overcome. However, while these expressions and quantities are invaluable, as they provide a value of resistance, they do not necessarily tell us how the flow of air behaves around the body.

2.5.2 Understanding Airflow Relationship Between Air Pressure and Air Speed

The flow around us when cycling causes a low-pressure region which creates an opposing force to motion (drag). This is caused by the fact that the energy of the flow needs to remain constant; so as the kinetic energy (velocity) of the air around the rider increases, it causes a pressure drop.

The nearest mechanical analogy is that of throwing a ball in the air, and as it gains height and potential energy, the speed reduces until it loses all of its kinetic energy, and as it falls, the opposite occurs. Hence, the drag across a moving body is very similar in terms of air pressure and kinetic energy (Eq. 2.10).

$$p + \frac{1}{2}\rho V^2 = \text{Constant} \tag{2.10}$$

$$\text{Pressure} + \frac{1}{2} \times \text{Density} \times \text{Speed}^2 = \text{Constant}$$

$$\text{Static Pressure} + \text{Dynamic Pressure} = \text{Constant}$$

Thus, in theory, if the velocity of air around a moving body remains constant, there is no reduction in pressure, thus no drag force. However, energy is also lost in the flow due to friction, which generates heat and sound. This means that the choppier the air, the greater the pressure deficit. Hence, this is one of the reasons fairings succeed in reducing the pressure drop, as they allow for a smooth air flow with no sharp transitions; thus, the velocity changes.

2.5.3 Air Boundaries

To understand why a stich might be added in a cyclist's body suit to trip the airflow around their shoulders to assist in drag reduction, it is worthwhile to explore the flow transitioning stages in general and to realise that a turbulent flow can be used to advantage.

As shown in Fig. 2.7, when a vehicle (such as a bus) is in motion, the flow initially slides around the side of the body in thin laminar layers. The laminar boundary layers stick to the body and move slowly due to the shear friction, it is like pulling a single A4 sheet of paper out of the packet.

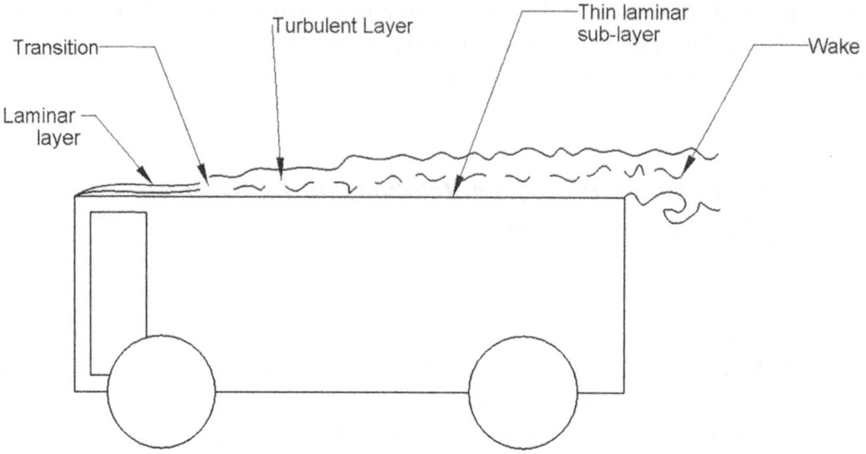

Fig. 2.7 Boundary layer effect. (Adapted from Barnard 2009, p. 9)

Fig. 2.8 Dust "sticking" to your bicycle after a dusty ride in areas of lamina flow

This explains why the dust never blows off the car as it is unfortunately stuck in the thin boundary layer next to the car body. This phenomenon can even be observed when cycling on a hot and dry dusty day (Fig. 2.8)!

The rest of the air is moving over this boundary layer. The surface of this laminar layer interacts with the fast-moving air and as a result grows outwards, increasing the thickness of the boundary layer. At a point depending on geometry, air viscosity and speed, the boundary layer transitions and becomes turbulent, after which, depending again on the aforementioned parameters, the flow eventually separates entirely. However, the turbulent air is still streamlined and follows the contours of a body more closely than a pure lamina flow. Indeed, that is why stitching is used in clothing to trip the air, so it becomes turbulent and better follows the body shape of the cyclist to delay separation and loss of energy; it is once the flow separates that the low-pressure region is formed.

2.5.4 *Air Moving Around a Cyclist*

On encountering a head or side wind, all cyclists ultimately know that to maintain speed they will have to expend a lot more effort. What is it about moving through the air even on a still day that causes resistance to motion? This is due to the surface friction and to a larger extent the pressure drag. As discussed in the previous section, the rearward facing surfaces experience a lower pressure than the forward-facing surfaces, resulting in a force that opposes the direction of motion (Barnard 2009). As shown in Fig. 2.9, flow separation occurs at the shoulder, and a low-pressure turbulent wake is formed almost immediately behind the cyclist (Crouch et al. 2017).

Figure 2.10 illustrates that the time trialist's aerodynamic position is effective, as it helps to reduce immediate air flow separation from the shoulders. Because when sitting upright, the air separates at the shoulders and has limited opportunity to follow the contours of our body—the cyclist is a bluff body. Additionally, it can be seen why a faired HPV reduces back pressure and can be pedalled at average speeds that exceed that of a conventional bicycle. Fairings provide a surface to ensure that the flow sticks to the length of the fairing, and as a consequence, the drag is primarily a function of the fluid shear forces. In 1939, the *Velocar*, a recumbent faired bicycle, was ridden 50.537 km at sea level and was the first human-powered vehicle to break the 50 km/h mark. Moser only achieved this on his bicycle in the 1970s at an altitude of 2300 m (Malizia and Blocken 2020).

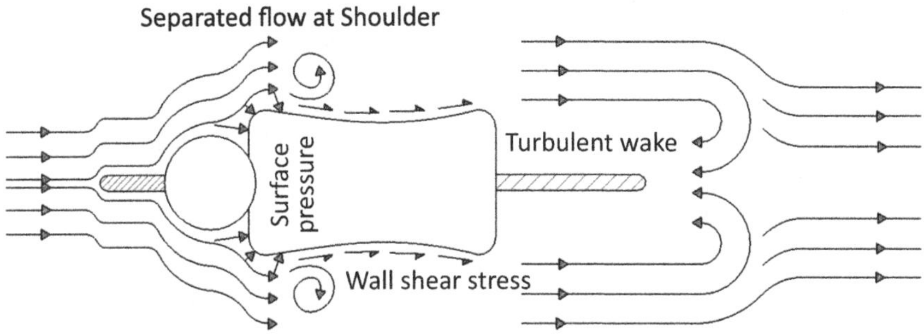

Fig. 2.9 Plan view of a simplified flow around cyclist. (Inspired by Martin et al. 2007)

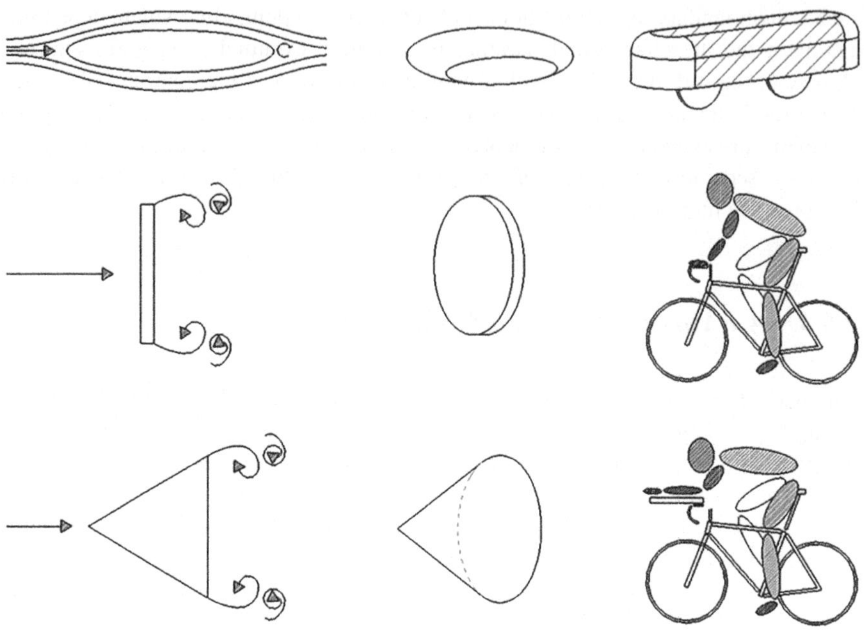

Fig. 2.10 Drag coefficients and analogous geometric shapes. (Inspired by Crouch et al. 2017)

However, during the pedal stoke, the body asymmetry which occurs when one leg is fully extended results in a worse back pressure, compared to when both knees are equally flexed, with the cranks both positioned horizontally (Crouch et al. 2017). As highlighted in Fig. 2.11, viewing airflow requires a three-dimensional perspective and a projected elevation or plan view as highlighted by Fig. 2.9 is limited. The low-pressure vortices created during the pedal stroke are very much like whirlwinds with low pressure and high speed, creating more back pressure. A significant amount of energy is used in creating such vortices, and as a consequence, very high drag forces are created.

This phenomenon is also seen in motorised forms of transport such as cars where the high pressure under the car is drawn upwards, which creates vortices. In comparison, when cycling, the motion of the extending and flexing thighs causes the drag force to vary by 20% through the pedal stoke, enabling high- and low-pressure flow to meet and create non-symmetric and symmetric vortices throughout the crank cycle (Crouch et al. 2017). It does not appear that crank length impacts these time-averaged drag components, and the largest potential impact on aerodynamic performance due to crank length is to open up the hip angle to allow for lower torso angles and more extreme aerodynamic positions (Wang et al. 2022).

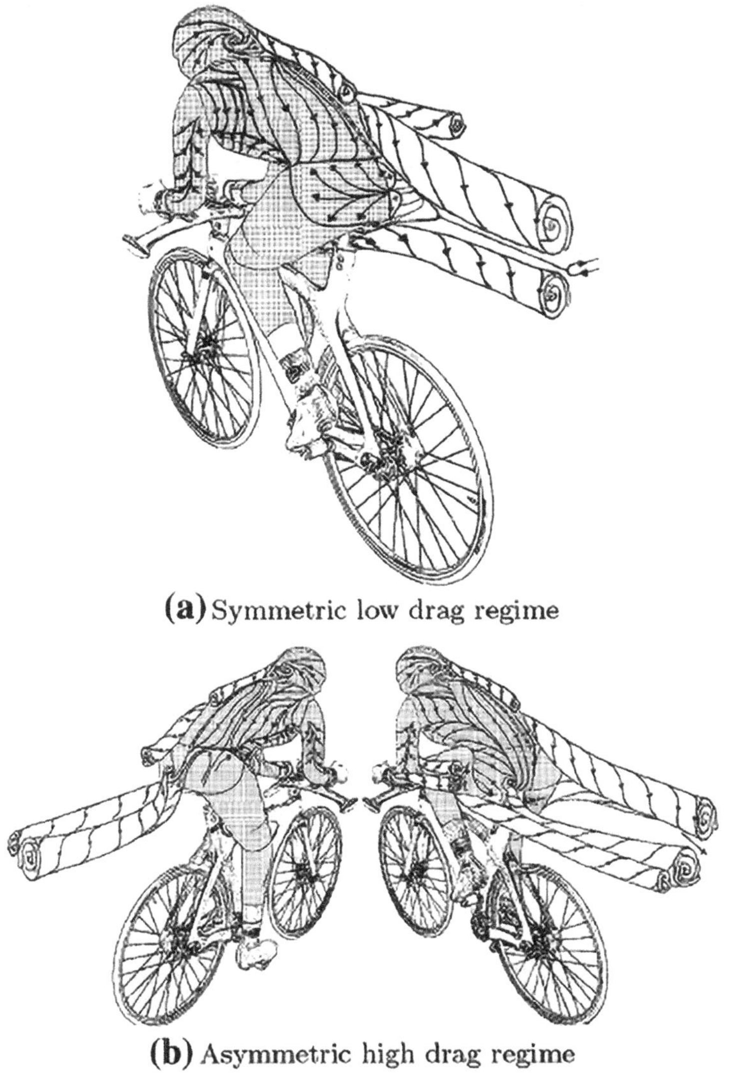

Fig. 2.11 Vortices created while pedalling. (Crouch et al. 2017)

2.5.5 Predicting Resistance to Flow

There are a few methods to evaluate the cyclist riding against the wind by wind tunnel analysis, analytical evaluation, computational fluid dynamics and finally by evaluating in the field. Each method has its merits although the greatest challenge faced is the continuity between results as studies may not be carried out to the same protocol or standard. Variation in local atmospheric conditions, such as pressure and temperature, affect fluid viscosity and air and can have a significant effect on the magnitudes of results (Barnard 2009; Crouch et al. 2017; Malizia and Blocken 2020). However, the results derived from both practical wind tunnel evaluation and Computational fluid dynamics (CFD) are invaluable for providing general rules of thumb and coefficients that can be used in analytical calculations. The small differences between new designs of wheels, frames and subtle changes in bicycle set-up are very difficult and nearly impossible measure. This is because even the slightest difference in experimental procedure causes a different result. This tells an important story in itself that aerodynamic components may only work for very specific conditions.

2.5.6 Analytical Evaluation

The advantage of analytical evaluation is that it is relatively easy to use dynamic equilibrium to estimate the relative gross effect that positional changes have on drag. Analytical evaluation does not provide a stand-alone solution even though it is possible to use simple known geometric shapes to approximate the profile of the cyclist position (Fig. 2.10). The coefficients are usually measured empirically by experimental wind tunnel analysis or estimated by CFD analysis; both these methods have limitations, and therefore analytical expressions cannot comment upon subtle geometric, clothing or positional factors. However, analytical analysis can provide a striking graphical summary of the power required to ride against the wind. For example, as shown in Fig. 2.12, the power requirement to ride against drag increases to the cube of the wind speed; the graph does not separate the effect of an oncoming headwind! Hence, if you were cycling happily at 24 km/h (15 mph) against a gentle 8 km/h (5 mph) headwind, this would have the same effect as cycling at 32 km/h (20 mph) and would require you to maintain an average 175-W output to maintain a steady speed. Figure 2.13 illustrates the effect, in general, how the different positions on the bicycle influence the drag force with the coefficients provided in Table 2.1. However, professional athletes will be adapted their technique considering their biomechanics. Therefore, do not adopt super aerodynamic positions and think you will go faster, as there is also a compromise between comfort and ergonomics/ flexibility to hold these positions.

2.6 Bicycle Frame and Wheel Drag

Incidentally, as shown in Fig. 2.14, the wheel rotates about the point of contact on the ground even though it is spinning about the hub: for the reason that the entire motion of a wheel is made up of the wheel spinning about the hub and translating in the direction of motion, hence highlighting the effect of evaluating a component like a wheel in a wind tunnel statically.

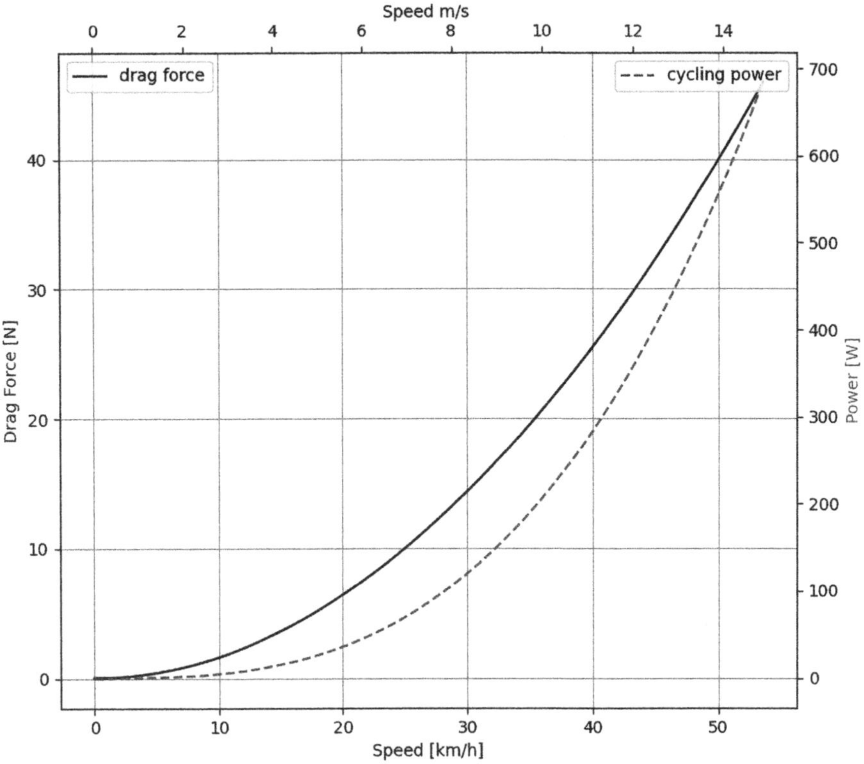

Fig. 2.12 Illustration of drag force, rider power output with respect to riding speed

At the time of writing, the final frontier appears to be the reduction of the drag force, and it is easy to appreciate why. There are numerous components on the bicycle ranging from the chain set to braking and electronic shifting to negate the use of cables so they do not cause drag. However, the idea of reducing drag is not new, and the concept of disk wheels was first patented in 1892 by Arthur Comings Hid. The wheels were manufactured by the Disk Wheel Company with claims of reducing air resistance. While history has shown us that they never became common place, it was clearly understood that the drag of wheels could be reduced, and that a complete front disk would cause instability, due to cross winds (Sharp 1896; Malizia and Blocken 2020) (Fig. 2.15).

Aerodynamic research carried out by the Royal Aircraft Establishment in 1919 documents the effect of covering the wheel spokes and using spats on aeroplane wheels to reduce drag at speeds of 30 m/s (67 mph). The investigations highlighted that the drag could be reduced by more than half (Zdravkovich 1992).

However, using a fairing to cover the spokes compromises wheel stability when there are cross winds on the bicycle. A full disk wheel is twice as prone to crosswind instability. The pressure centre acting on a disk wheel compared with spoked or

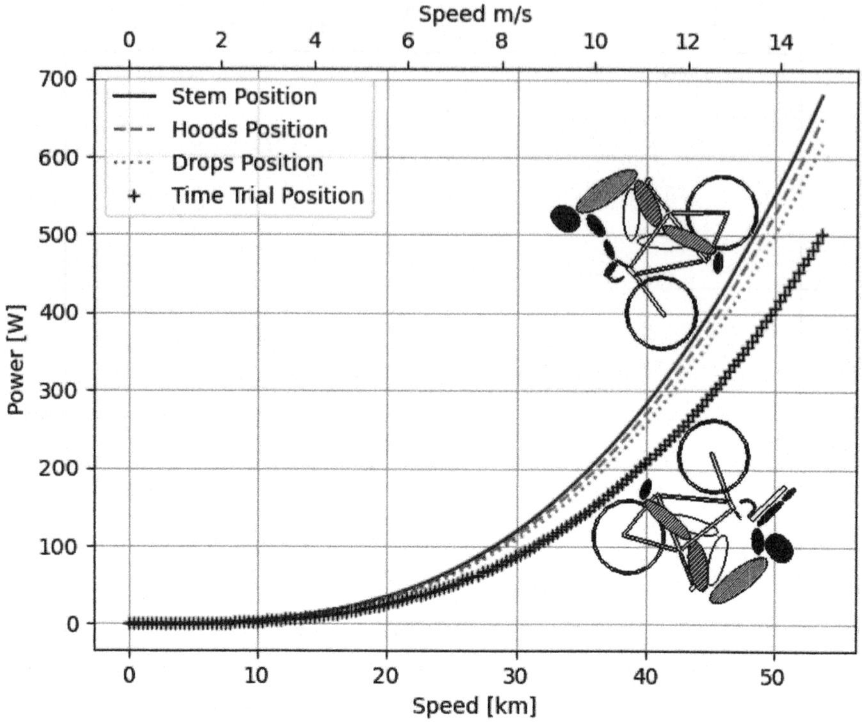

Fig. 2.13 Prediction of cyclist power output for drag using values in Table 2.1

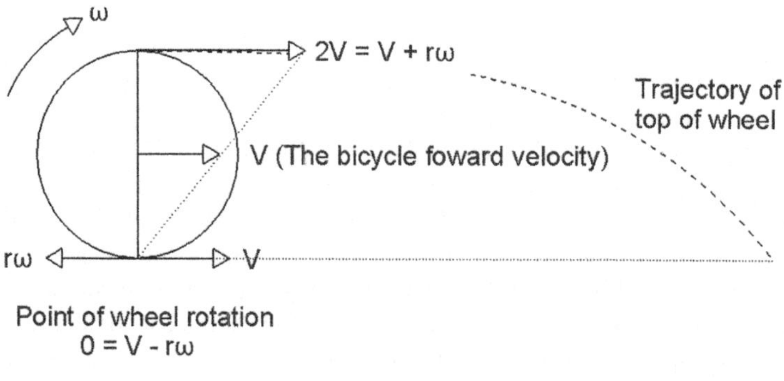

Fig. 2.14 The vector schematic of a rotating wheel

tri-spoked wheels is, respectively, located 0.25 and 0.13 times the diameter of the wheel axle from the centre.

In 1984, when Moser broke Merckx's record with (Lenticular) disk wheels, the event heralded a renewed interest aerodynamics (Malizia and Blocken 2020), especially considering that Merckx's team focused on building the lightest possible bicycle while Moser's team built the bicycle with the lowest possible drag resistance

Fig. 2.15 Initial disk wheel concept. (Sharp 1896, p. 352)

(Monte et al. 1987). However, since Moser broke the hour record, there have been numerous studies on the effect of wheel drag. Malizia and Blocken (2020) to date has complied a fairly extensive summary of both CFD simulation and wind tunnel evaluation studies, such as the "traditional" 36-spoked wheels, aerodynamically spoked wheels (tri- and four-spoked) and disk wheels. In summary, while it is not possible to provide objective values due to the difficulty of comparing results across different studies, useful "rules of thumb" when considering the aerodynamics of wheels are as follows:

- The lenticular and disk wheels showed the lowest drag resistance with a 16% maximum difference.
- The best performing aerodynamically tri- and four-spoked wheels showed a greater deviation in performance of 38%—though there were some that performed well as the disk wheels.
- The spoked wheels in general have greater drag resistance compared to disk wheels, though the results were sensitive to rim profile and the number of spokes with the results showing a 50–150% variation compared to disk wheels.
- The shape factor of the wheel: The ratio of the wheel/tyre diameter can be used to predict the influence of the drag coefficient: with increasing tyre diameter, the shape factor decreases and so does the drag resistance (Zdravkovich 1992).

Clearly, there is a great deal of technical nuances to consider. The airflow around any rotating wheel is highly sensitive to the tyre and wheel combination, as well as small differences in tyre and wheel geometry. Thus, many studies fail to find a good deal of correlation between the drag and lift coefficients, determined from either CFD or wind tunnel studies. In general, as illustrated by Fig. 2.16, the rotating wheel produces a highly separated flow with a broad wake and vortices which in turn results in wheels having a high drag. Additionally, the spinning wheel throws the air

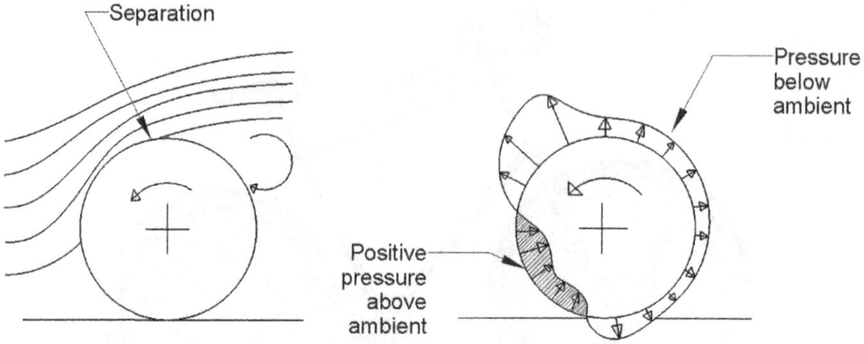

Fig. 2.16 Pressure distribution and streamline patterns for rotating wheel. (Adapted from Barnard 2009, p. 37)

forward into the direction of travel, especially at the top of a wheel Fig. 2.14, traveling at twice the instantaneous speed in the direction of travel (Barnard 2009).

2.7 Conclusion

The purpose of this chapter is to provide the reader with an understanding of the forces that act on the bicycle rider. It is surprising that only 5–20% of the pedal force is transmitted to the rear wheel. Hence, it is unsurprising that small changes in resistive forces caused ether by mechanical losses, drag or gravity can result in significant speed reduction. However, despite drag being the holy grail at the time of writing (and it is easy to see why), the significant reduction in drag claimed by manufacturers of bicycles and their components is unlikely to be translated into that experienced by the everyday cyclist. These components work most efficiently at the racing speeds of professional athletes, and they are often adapted using 3D printing technologies for the rider in question. Additionally, components that are wind tunnel evaluated with significant watt-saving claims without fitment to a bicycle, or fitted to a bicycle without a rider, will not realise such benefits in practice.

Since the first Tour de France in 1903, designed to promote the latest technology and sell bicycles, advancements in technology and the concepts behind the adaptations are in themselves fascinating to witness, although they at time distract from the biggest influence on the bicycle.

There are many tools available for the cyclist enthusiast that can be used for training and improving their physicality. Although, it does not always seem that they are used to best effect. For example, there are a number of reviews on power metres that compare the time average cadence power to highlight which one deviates from the average. While studies are informative, such information does not help you improve your riding technique. They are not designed to be anything more than indicative and set a scale for power cadence training. The tools used by cyclists to set thresholds or use on static trainers can ensure that the bicycle is set-up to meet your body anthropometrics and biomechanics. It is on that note we conclude this chapter and progress to the chapter reviewing cycling biomechanics.

Bibliography

Barnard RH (2009) Road vehicle aerodynamic design: an introduction. MechAero Publishing, Albans

Burke E (1986) Science of cycling. Human Kinetics Publishers, Champaign

Crouch TN, Burton D, LaBry ZA, Blair KB (2017) Riding against the wind: a review of competition cycling aerodynamics. Sports Eng 20(2):81–110

Gibertini G, Grassi D (2008) Cycling aerodynamics. In: Nørstrud H (ed) Sport aerodynamics. Springer, Vienna, pp 23–47

Malizia F, Blocken B (2020) Bicycle aerodynamics: history, state-of-the-art and future perspectives. J Wind Eng Ind Aerodyn 200:104134

Martin JC, Davidson CJ, Pardyjak ER (2007) Understanding sprint-cycling performance: the integration of muscle power, resistance, and modeling. Int J Sports Physiol Perform 2(1):5–21

Monte A, Leonardi LM, Menchinelli C, Marini C (1987) A new bicycle design based on biomechanics and advanced technology. Int J Sport Biomech 3:287–292

Sharp, A. (1896). Bicycles & tricycles, an elementary treatise on their design and construction. Lond. &, The MIT Press

Smethurst AJ, Findlay IS (2023) Has 'aero is everything' gone too far? J Sci Cycling 12(3):4–9

Sneiderman, P. (1999). Pedal power probe shows bicycles waste little energy. https://pages.jh.edu/gazette/1999/aug3099/30pedal.html?form=MG0AV3

Wang S, Pitman J, Brown C, Tudball Smith D, Crouch T, Thompson MC, Burton D (2022) The influence of the inter-relationship of leg position and riding posture on cycling aerodynamics. Fluids 7(1):18

Zdravkovich MM (1992) Aerodynamics of bicycle wheel and frame. J Wind Eng Ind Aerodyn 40(1):55–70

Chapter 3
Biomechanics of Cycling

3.1 Introduction

Essentially, a well set-up basic bicycle will enable the cyclist to go further and faster more comfortably than the latest mechanical or aerodynamic technology bicycle. A correctly positioned engine is more important than any other aspect of technology on the bicycle.

The field of biomechanics uses a variety of methods and tools: from observations by trained clinicians, specialist laboratories quantifying the motion of the human body and equipment in the field, for example, by utilising in-built dynamometers in the case of cycling. The terms used when quantifying the motion (kinesiology) of the human body are the kinetics and kinematics of motion, respectively, relating to the muscle forces that cause that motion and outcome of that motion by observing and quantifying limb positions. These techniques and how they are used along with observation and user experience to maximise the rider's power output will be further discussed in the following chapter.

At the time of writing, more freeware tools are available than ever before to study and evaluate motion, be it cycling or other sport activities. There are even affordable mobile phone application tools which utilise artificial intelligence (AI) to help any rider set-up their bicycle with improved precision. Therefore, the purpose of this chapter is to provide the reader with an insight into biomechanics and the type of measures that are generally inspected. Furthermore, a better understanding of this area will assist the reader to decipher the vast swathes of newsfeed that today's cyclist is infiltrated with.

Even though biomechanics provide us with useful measures to, for example, set the saddle height so there is a minimum knee flexion of 35° or provide us with an insight in asymmetries of power developed between the dominant limb, biomechanics do not tell us why or how the motion is created. Therefore, in the next chapter (Chap. 4), an understanding of how muscles function and how force is transmitted

A. Crimin, A. McGarry, *Bicycle Biomechanics*,
https://doi.org/10.1007/978-3-031-85712-6_3

through the body is also presented, otherwise the kinematics and kinetics may be misinterpreted or not thoroughly challenged.

3.2 Anatomical Terms

Before building a fuller picture of how the muscles transfer power across the limbs and the different types of muscle contraction, it is worth highlighting some common terms used to describe limb motion (Fig. 3.1). Additionally, while the majority of limb motion occurs in the sagittal plane (viewed from the side), there is also motion in the coronal (frontal) and transverse planes, most notably hip adduction and abduction (Fig. 3.2). For the purpose of this book, however, the discussion will primarily be focused on the sagittal plane.

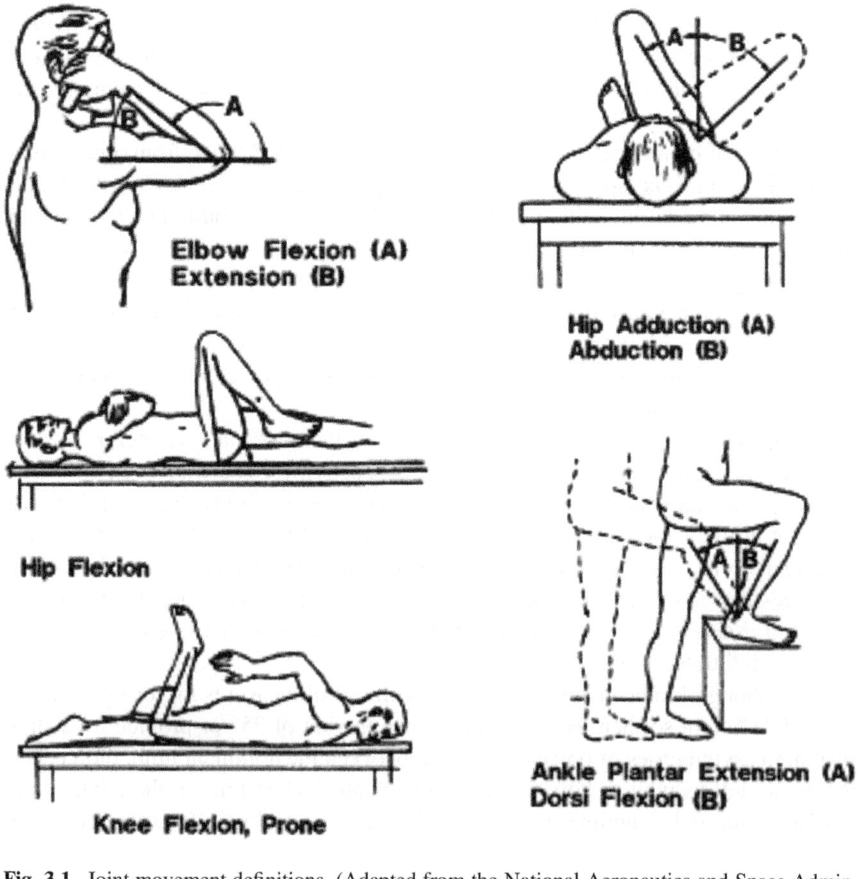

Fig. 3.1 Joint movement definitions. (Adapted from the National Aeronautics and Space Administration (NASA) website (Thornton and Jackson 1980))

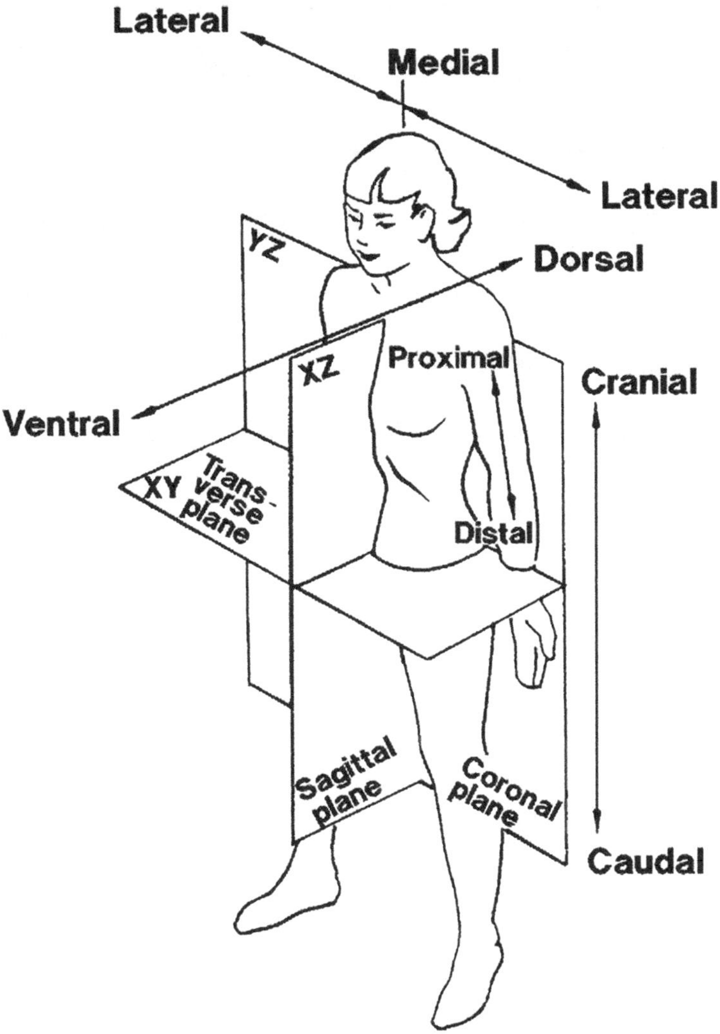

Fig. 3.2 Anatomical planes from the National Aeronautics and Space Administration (NASA) website (Thornton and Jackson 1980)

3.3 Biomechanics

One of the easiest to use evaluation tools that we have at our disposal to compare the motion of individual with themselves (intra-subject) or that with others (inter-subject) is simple observation. Indeed, simply observation can be achieved by taking images and observing the motion through a lens. Leland Stanford commissioned Eadweard Muybridge to take photos of his champion race horse to settle the debate

Fig. 3.3 Muybridge's *The Horse in Motion*, 1878

as to whether all 4 ft of a trotting horse are ever off the ground at the same time (Fig. 3.3).

Even though trained observers and video camera evidence allow for the subtleties of motion to be quantified by visual observation such as the effect saddle position has on the rider pushing their pedals, it is difficult to objectively quantify the push and its effect without an objective measurement. Biomechanics is the branch of science focused on the technique of objectively quantifying human motion or indeed that of any living animal. A number of multidisciplinary techniques are used in this field to either evaluate the effect on motion or capture the movement (kinematics) and the forces (kinetics) of motion. Tables 3.1 and 3.2 summarises methods employed to measure the kinetics (forces) and kinematics (trajectories) of motion such as video evidence, inertial sensors, force gauges, pressure mapping systems and even electromyography (EMG) (Millour et al. 2023).

However, the aspect of biomechanics that is often overlooked is that of muscle physiology. Cycling is a repetitive activity predominantly employing the muscles of our lower limbs, requiring the repository system to support our musculature activity. However, what contraction factors influence our pedal rate and power that is generated during a crank revolution? How does the saddle height effect the muscle extension length and the force generated? Does the position that we instinctively find best suit our riding style; is there anything about muscle physiology that tells us that we can do it better? To that end, the following chapter will provide an introduction to muscle physiology.

Table 3.1 A summary of techniques that can be used for kinetic analysis

Tool	Method	Validity and reliability
External force plate measurement	Captures the external ground reaction force	Good accuracy and can capture detailed patterns of force
Force gauge measurement	Cannot be inserted into a living person, though can be fitted to a device such as a component interacting with a person and used to measure the reaction force	Allows for measurement both in laboratory and in-field
Pressure mapping	Electronic film or passive ink film can be used to measure the pressure distribution such as a saddle interface	Can be used to tell that the biological interface is loading as expected
Mechanical modelling	Similar to kinematic modelling but can also be used to determine the forces that cause motion	As in kinematic modelling, information can be obtained by fixing variables, though information is only as good as model
Inverse mechanics	This sort of analysis usually involves capturing kinematics and kinetics of motion, which is used to determine the positions and magnitudes of forces around joints	Can be used to summarise the gross effect of muscle usage

Table 3.2 A summary of techniques that can be used for kinematic analysis

Tool	Method	Validity and reliability
Video annotation	Recorded motion of interest using a standard video recorder.	Such video motion capture tools are readily available in the form of freeware packages such as Kinovea, making kinematic analysis readily available
3D motion capture systems	Retro-reflective markers are located on anatomical prominences and used to determine an underlying rigid structure which can be used to calculate joint motion, thus angles in 3D	This sort of analysis is still considered the gold standard of kinematic analysis though not readily available
Mechanical modelling	Kinematic modelling allows the motion of joint to be determined mathematically (p. 62)	Is not prone to test protocol, motion capture hardware and software limitations, though modelling is only as good as the input when predicting outcomes
Indexing	The technique, for example, of walking in sand and using a measuring tape to measure stride length	Easy to implement but not applicable to all motion measurement
Inertial measurement unit	The use of small sensors with accelerometers and/or gyroscopes that can be used to capture the acceleration or orientation of limbs	Allows for measurements both in laboratory and in-field

3.4 Cycling Kinematic and Kinetic Analysis

3.4.1 Kinematic Analysis

To kinematically asses the trajectories of the lower limbs, the pelvis, hips, knees, ankles and foot metatarsal positions are measured with respect to a coordinate system. Markers are usually placed on identifiable bony prominences such as knee epicondyles, ankle malleolus and the iliac crest of the pelvis—for both 2D video analysis and full 3D motion capture. For 3D analysis, the markers are usually spherical balls with a flat base and are covered in glass beads that are approximately 10–14 mm in diameter. The reason spherical markers are preferred is that their centre is more easily returned: there is more error in returning the position of a flat reflective "sticker". Statistical formulae are then used to identify hip, knee and ankle centres from the position of bony prominences, and thus allowing a rigid link to be drawn, for example, for the legs: from the hip to knee, knee to ankle and finally to the foot. Figure 3.4 provides a visual insight into how the external markers are placed over anatomical prominences and used to build a rigid structure of the body. For 2D analysis, it is usually assumed the joint centre lies directly below the marker. Furthermore, it is worth noting that for 3D analysis, the location of joint centres is not always estimated to be directly below the markers as in 2D analysis. However, while 3D analysis is invaluable, it is not the most accessible tool and is not necessarily required when investigating motions that are already published in literature such as walking or cycling.

An important variable that can be determined from kinematic analysis is the magnitude of limb flexion and extension angles. 3D kinematic analysis does allow us to measure limb flexion angles more accurately. Compared with a 2D video image, the angle magnitude (accuracy) can be affected depending on how the camera is placed (angled) with respect to the motion being filmed. For example, if you shine a

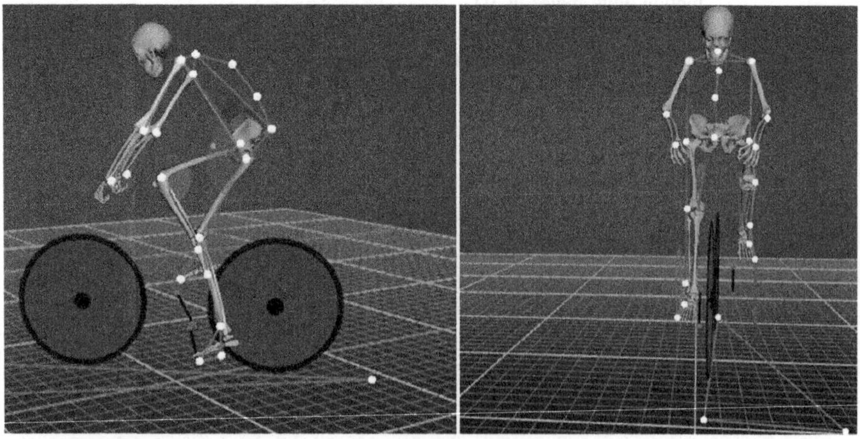

Fig. 3.4 Rigid body links determined from external markers. (Millour et al. 2023)

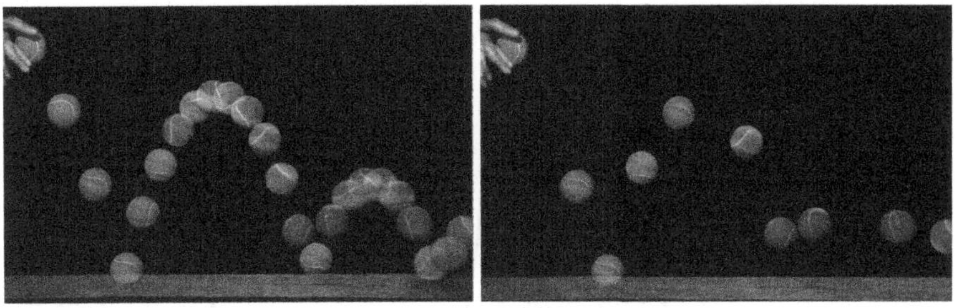

Fig. 3.5 Multiple-strobe shot of a bouncing tennis ball at 20 Hz (left) and 10 Hz (right). (Image courtesy of Terry Oakley/Alamy Stock Photo)

torch against a 30 cm ruler and cast shadow against a wall, the shadow length will differ from the true length unless the light source is truly normal; this is where 3D analysis can provide improved accuracy.

Another challenge of capturing the trajectories of motion is due to the limited capture frequencies of cameras in general, caused by the physical limitation of opening and closing the shutter lens. Thus, it is hard to capture motion accurately above 60 Hz if the maximum capture frequency is <120 Hz; this is known as the Nyquist frequency, and the sampling rate should be at least double the frequency being captured. Though in reality, it should be greater as electronic noise and other hardware/software factors come into play. As highlighted by the strobe images of the tennis ball, it is difficult to determine the parabola trajectory of the tennis ball when the capture frequency is halved (Fig. 3.5).

For the purpose of this book, to show how the everyday cyclist can explore their biomechanics to enable them to improve their technique and bicycle set-up, the open-source kinematic software Kinovea (www.kinovea.org) is utilised. Kinovea has been shown to have good accuracy when analysing motion in the sagittal plane, with discrepancies of 2–3° compared to commercially available software (Millour et al. 2023). However, it is the opinion of the author having used "state-of-the-art" 3D systems such as *Vicon* that a magnitude difference of 2–3° can result from positioning markers on bony prominences, as these need to be located by palpating the joint through the skin which is prone to interpretation.

Setting up a camera in a room at home or in the garage to objectively investigate ankle, knee, hip and trunk angles for different set-ups is an invaluable tool. There are numerous readily available *Kinovea* tutorials online, and the analysis shown below can be carried out by watching introductory 10-min tutorials. Figure 3.7 shows the author using the analysis to measure his kinematics while using a turbo trainer; the image also illustrates the definition of the limb flexion angles. Figure 3.9 plots the angles that are typically published in literature such as Carpes et al. (2014) for the hip, knee and ankle. Despite International Society of Biomechanics (ISB) recommending standards (Wu G 2002), with which to report such angles, the reader will find different reported angles than shown here, if that is the case, then check the orientation definition. In general, it should be the orientation of the distal segment

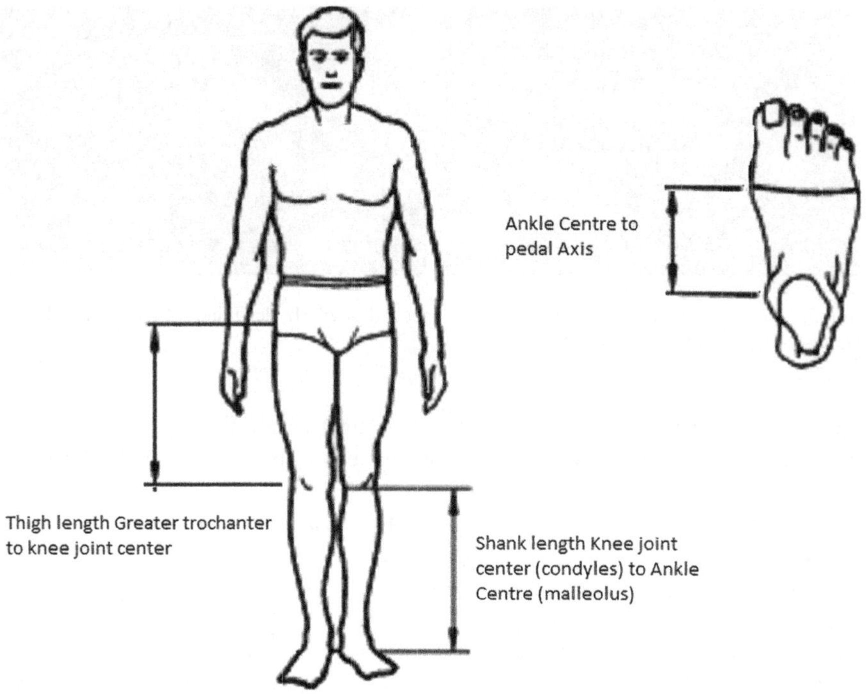

Fig. 3.6 Anthropometric parameters required for kinematic analysis. (Adapted from the National Aeronautics and Space Administration (NASA) website Thornton and Jackson 1980 and definitions from Merrill et al. 2019)

with respect to the proximal segment; that is the knee angle is measured by the difference of the angle between the shank and thigh because the thigh is also moving. If the knee angle was measured by a vertical line drawn in the image, the shape of the plot would not compare very well between different individuals, making it hard to plot general motion curves.

It should be noted that the foot segment shown in Fig. 3.7, running from the ankle to the pedal, is not an anatomical but a mechanical rigid mechanical link; highlighting the mechanical links do not necessarily tie in with the skeletal structure. However, a marker on top of the fifth metatarsal can be used to also define a foot segment. The segment lengths, as illustrated in Fig. 3.6, can be measured directly or using the *Kinovea* software. The author prefers the latter because white stickers were placed over the palpated anatomical prominences, and it is easier than using a measuring tape especially if you are conducting the measurement yourself! (Fig. 3.7)

The angles plotted in Fig. 3.9, respectively, show that the hip is extending until bottom dead centre (BDC) before flexing again; the knee is extending with a minimum flexion angle of approximately 30° at BDC before extending again, and the ankle at top dead centre (TDC) is still plantarflexed and becomes more plantarflexed at BDC. The limb kinematics are useful to view in this way as they

Fig. 3.7 Kinematic analysis of the right leg

can be compared to that published in literature, and it becomes evident that the limb motion is reasonably typical. For example, studies such as Nordeen-Snyder (1977) used the "smoothness" of the ankle kinematics to comment on when the saddle was position too high. However, to investigate how the muscles function or move the joint, the kinetics of motion also need to be investigated (Figs. 3.8 and 3.9).

3.4.2 Kinetic Analysis

The measurement of forces may be done in conjunction with or without the kinematic data. This could be simple as putting a load measurement device in products such as a vacuum cleaner handle to measure the pull experienced on certain surfaces, the handle of a tennis racket or in a prosthetic leg of a person with amputation to measure the ground reaction force. Thus, the advantage that cyclists have with respect to other sports/activities is evident: because of the manner in which their limb motion is used to create drive, it gives opportunity to measure the summation of all the effort by use of a power dynamometer. Hence, it is possible to assess individual limb kinetics and address asymmetrical limb effort (by use of individual pedal or crank meters), which in turn can be used to point to weakness in certain muscle groups.

However, it can be seen that it is impossible with the current technology to measure the loading around biological joints of limbs directly. Therefore, unless some sort of kinematic analysis is carried out in conjunction with the load measurement, it is not possible to determine the forces and moment (of the force) developed by the muscles around joints. This is why in biomechanics, the measurement affects

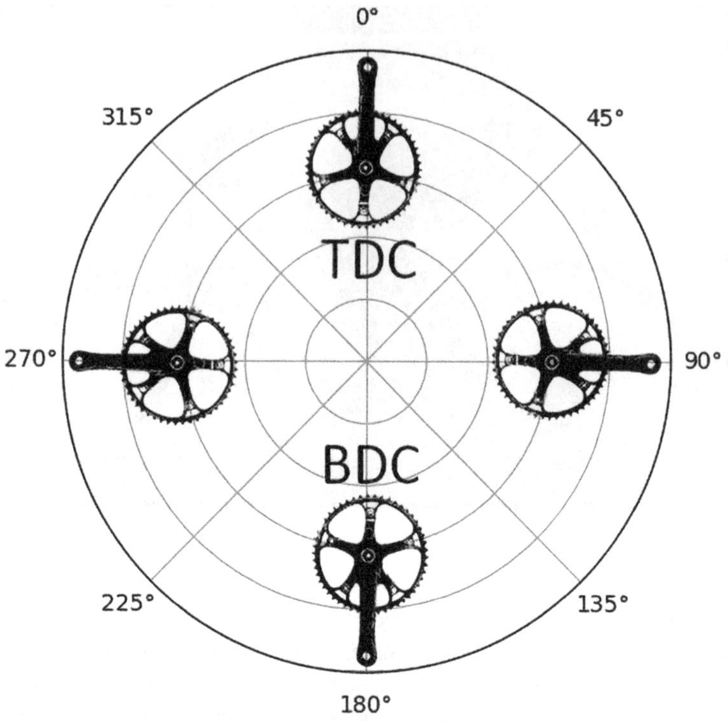

Fig. 3.8 Pedal stoke definition angular position (θ), 0° top dead centre (TDC), 180° bottom dead centre (BDC)

the activity. For example, sitting on a static bicycle while either being filmed is only reflective of how the activity is normally carried out on the road.

Figure 3.13 (p. 56) illustrates the 2D trajectory trace of the hip, knee and ankle for the cycling on the static bicycle shown in Fig. 3.7 (p.51). The crank forces measured at the instance shown in Fig. 3.13, include the effective/tangential force normal to the crank and the axial/radial force parallel with the long crank axis. It is the effective and radial force that most dynamometer systems will measure directly, rather than the applied pedal force. For the reason that the effective force is always in the same orientation with respect to the crank systems as it rotates, it is actually measured in a crank-based local coordinate system. To be useful and allow us to evaluate the moment of the force around the muscle joints, the force needs to be viewed with respect to the global coordinate system—that is along the horizontal and vertical edges of the image. For example, if you were asked to measure the length and breadth of a book, it is natural to measure along the orthogonal edges, as they are essentially the x and y direction of the book coordinate system. If the book were placed on a table at an angle shown in Fig. 3.10, the projected length and breadth to the table edges would be different to the true distance measured along the book edges; to align the book mathematically, a rotation is required.

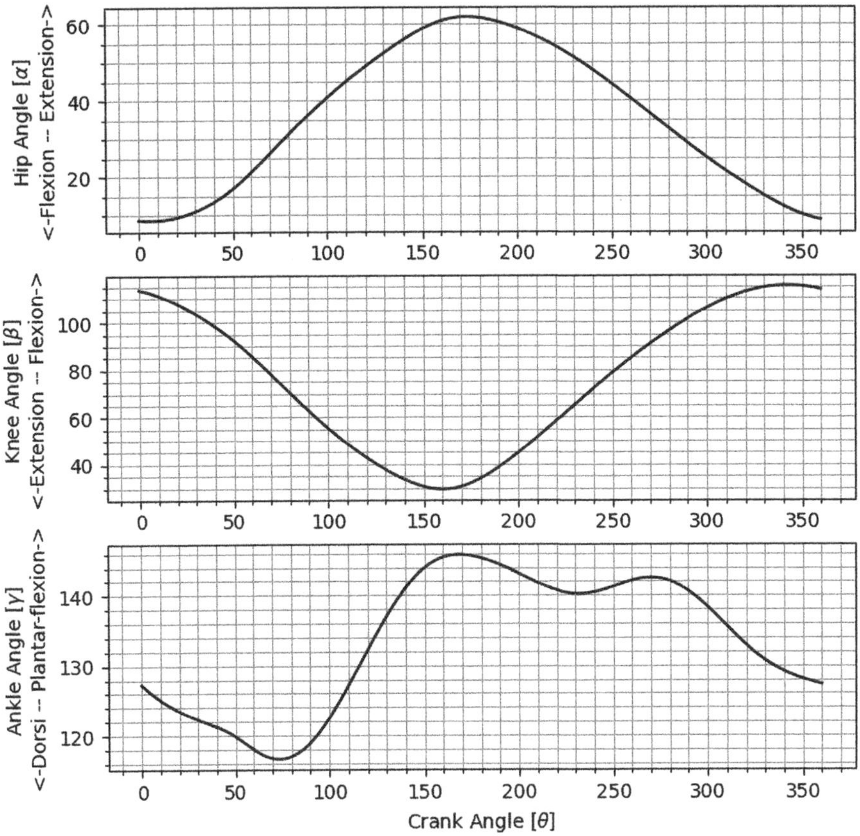

Fig. 3.9 Hip, knee and ankle angles measured using Kinovea

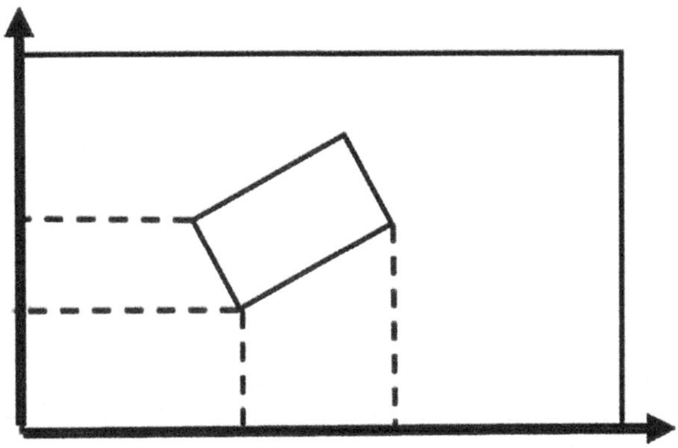

Fig. 3.10 Measuring edges of a book

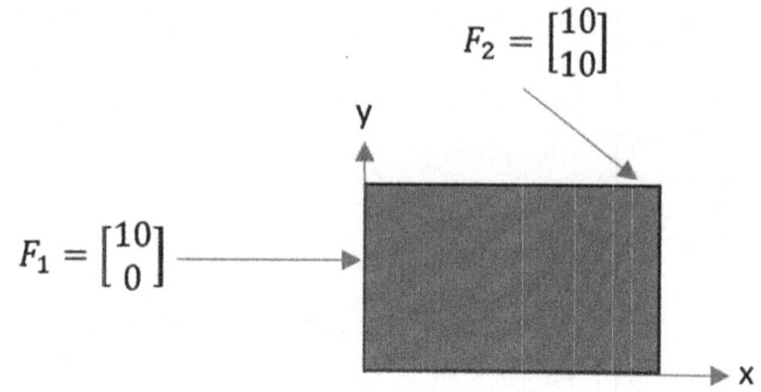

Fig. 3.11 Forces measured in a Global Reference Frame

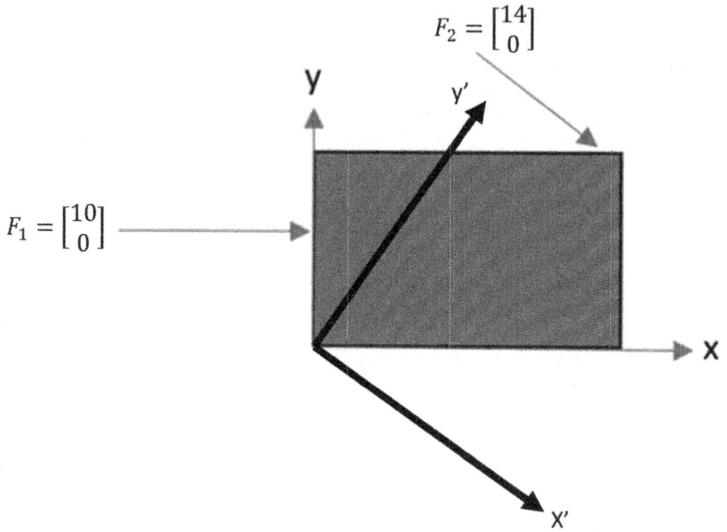

Fig. 3.12 Force measured in a global and local reference frame

Furthermore, like measuring the area of a rectangle, two lengths are required, to define a force (in 2D) in two components, measured in a global frame of reference (Fig. 3.11). For example, the horizontal force F_1 does not need to be written in vector matrix notion because its direction is evident. However, the orientation of force F_2 is not obvious on inspection, so the x and y components are required so the direction of the force is known.

The F_2 force (Fig. 3.12) is now measured in a rotated (local) reference frame because it was more convenient physically to do so—like the rotating crank. The force is exactly the same as shown in Fig. 3.11, and the x' axis is parallel to its direction giving rise to the 0 y' component. Nonetheless, to carry out calculations such as the turning effect, the force has around a point; it needs to be rotated from the x', y' axis to the x, y axis. Hence, in the same manner, the measured pedal, crank force, hub or bottom bracket forces are measured in a local measurement system; they need to be rotated/resolved so muscle moments can be determined. The reader does not need to understand any more than this, although it is useful to have an awareness that measurements are made with respect to a frame of reference even if this is done intuitively when taking the dimensions of a table surface.

Once the measured crank force is rotated and known in global coordinates, the muscle pulls around a joint can be calculated; this is shown schematically in Fig. 3.13. The dashed F_{global} pedal force magnitude can be projected by scale to estimate the perpendicular distance about the point of rotation and calculate the effect of a muscle (moment) pull (Eq. 3.1).

$$\text{Muscle moment } (M) = F_{\text{applied}} d \tag{3.1}$$

Once the moment around the crank rotating axis or the ankle, knee and hip are known, their power may be calculated by taking the product of the moment with the mechanical or biological joint angular velocity (ω).

$$\text{Power } (P) = M\omega \tag{3.2}$$

Hence, power dynamometers actually measure force rather than power directly. Figure 3.14 shows an example of the measured effective/tangential pedal force and calculated power. It can be clearly seen that the force output varies sinusoidally, although many power meters return the time averaged series data of a pedal revolution, which is better for power cadence. There are commercially available of-the-shelve systems, such as *SRM by* a German-based company or the Australian-based *INFOCRANK* that provide a full force plot for a pedal revolution. There are considerable challenges to building crank or pedal-based systems, not only with calibration but also with the electric hardware. In the final chapter, the challenge of building such a device is presented by practical demonstration.

There are a number of noteworthy aspects of this simple plot which can reveal interesting details about a rider's style, and in this case, the author's style at the time, the data was plotted after a road traffic accident. First of all, there is clearly an asymmetrical power output when comparing the right and left leg, with a respective average of 238 and 316 W for the left and right leg, meaning that there is a 25% difference—and you do not need statistics to tell you that is significant! Also, there are nuances in the pedal stoke that should be observed; the peak values of the right limb are greater, while the in-phase trough values of the left leg are more negative than the right. This, in fact, reveals that the right leg is contributing a significant effort to lifting the left leg up in the upstroke—during the recovery phase. Because to record a negative power during the upstroke, the left leg will have applied a downwards (negative) force during this period, opposing the motion of the crank—work is done against the crank. Conversely, it can be intuitively reasoned that when pushing down on the pedals, when the force and the angular velocity of the crank is in the same direction, positive work (power) is done on the crank.

The calculated anatomical powers for the captured data are provided in Fig. 3.15 for a single crank revolution from TDC to TDC, the curves are typical of what are

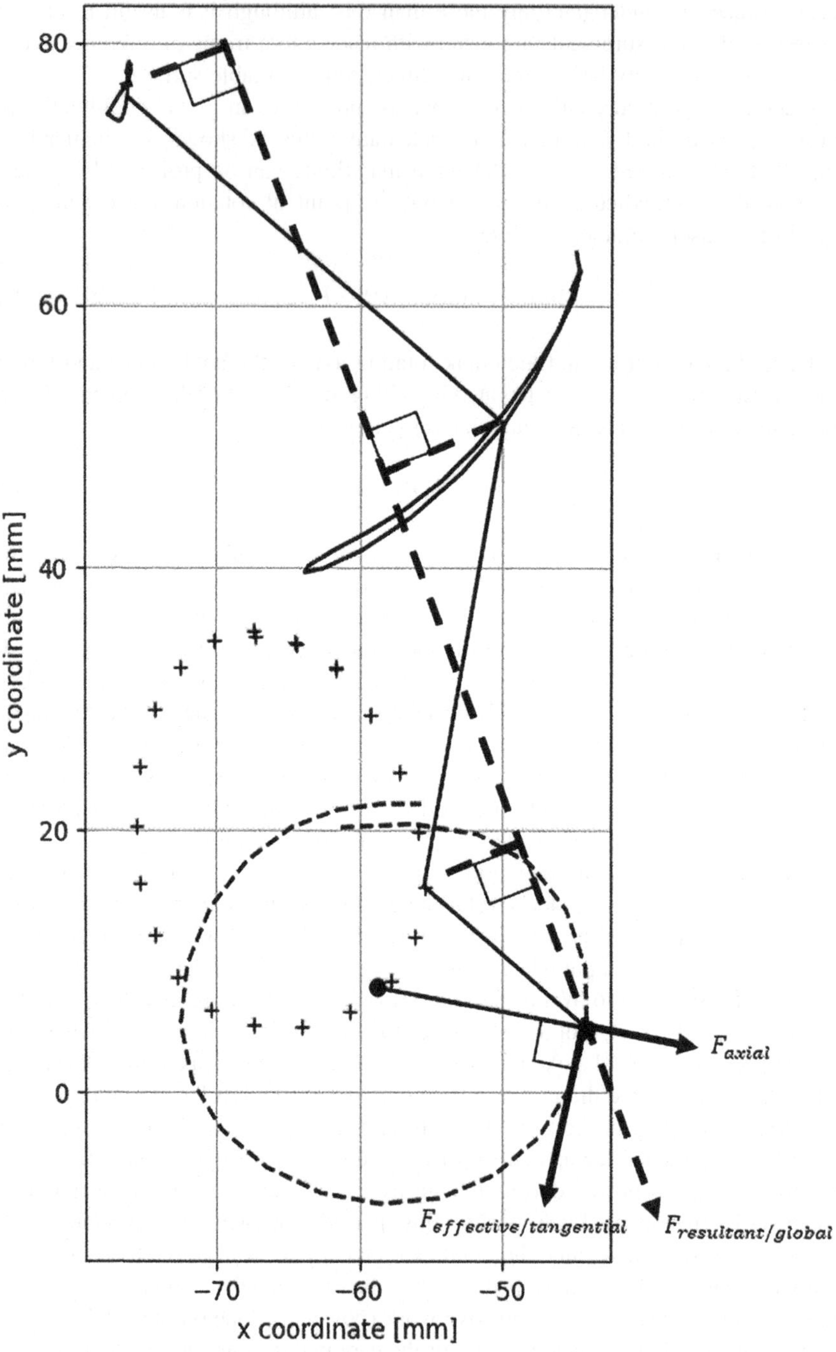

Fig. 3.13 The mechanical system used to determine lower limb kinematics derived from the Kinovea kinematics Fig. 3.7

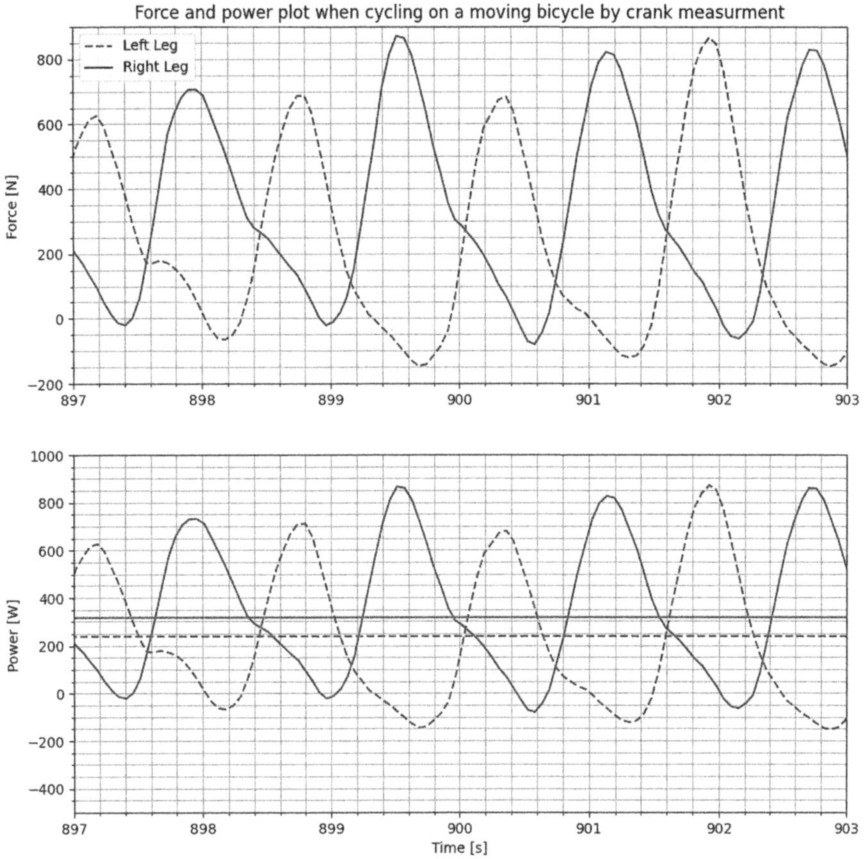

Fig. 3.14 Measured force (top) and calculated power (bottom) of left and right pedal stoke. Average power shown as straight lines

generally published in literature such as Barratt et al. (2016) and Elmer et al. (2011). They show that the main muscle power delivery to the crank occurs after the crank position of 90 degrees. The hip and knee musculature are responsible, in respective order, for the primary power developed in the downstroke, while the ankle musculature acts to transfer power from the thigh and shank, evidenced by the fact its peak occurs after that of the maximum hip and knee power. The developed knee power in the upstroke from BDC to TDC appears to be to help lift the knee, minimising the work required by the opposing limb during the downstroke.

By plotting the effective tangential crank force, the timing of maximum force development by the rider is clear—and is not at the extremes of maximum leg flexion or extension, respectively, at TDC or BDC. This in itself suggests that adjusting the saddle height does not directly affect muscle power generation but influences it by giving the muscles time to generate force and time to recover on the upstroke. If the maximum force development/power production occurred at BDC, incorrect saddle

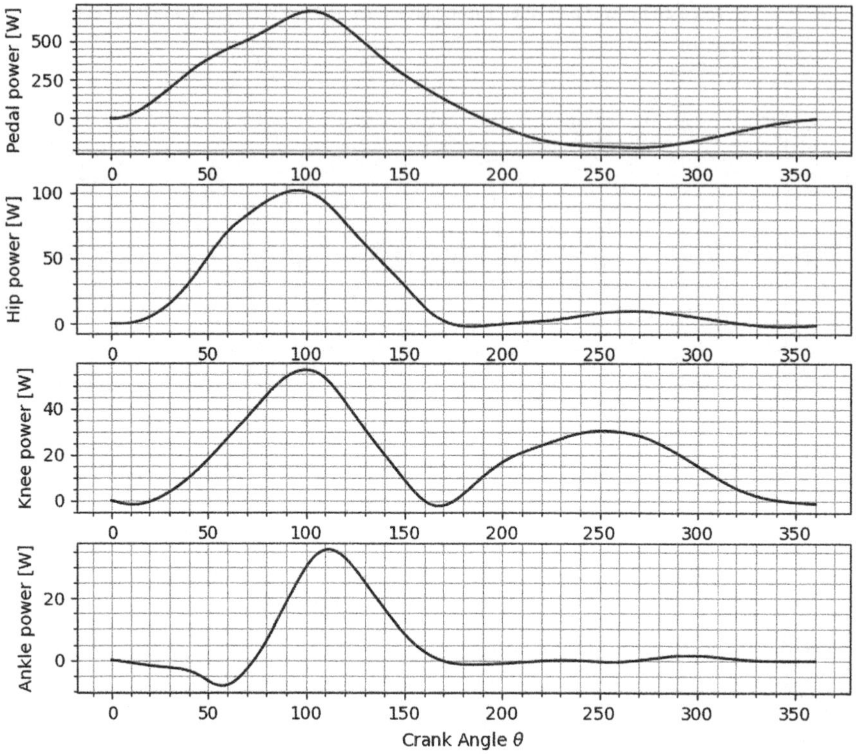

Fig. 3.15 The hip, knee and ankle powers when cycling

height would have a definite and critical effect on joint power production—which is not the case. There is a tolerance to the adjustment and how the rider can position themselves on their bicycle. However, a pertinent point to take is not to try to pull up on the pedals; it is not observed by professional cyclists, and the upstroke is required to help the muscle recover (Silberman et al. 2005).

The vector pedal shown in Fig. 3.16 is good for visualising, where in the pedal revolution, the main pedal force in the downstroke is applied. Furthermore, that at the bottom of the pedal stoke, the foot may pull back, but there is minimal effort applied to pull the pedal during the upstroke.

3.5 Mechanical Modelling

This section discusses how a five-bar mechanism model shown in Fig. 3.19 may be used to analyse the biomechanics of your cycling motion out with the confines of a laboratory. This sort of model may be used by bicycle fitting applications and gives the reader an insight into how our limbs move during a pedal revolution. There are

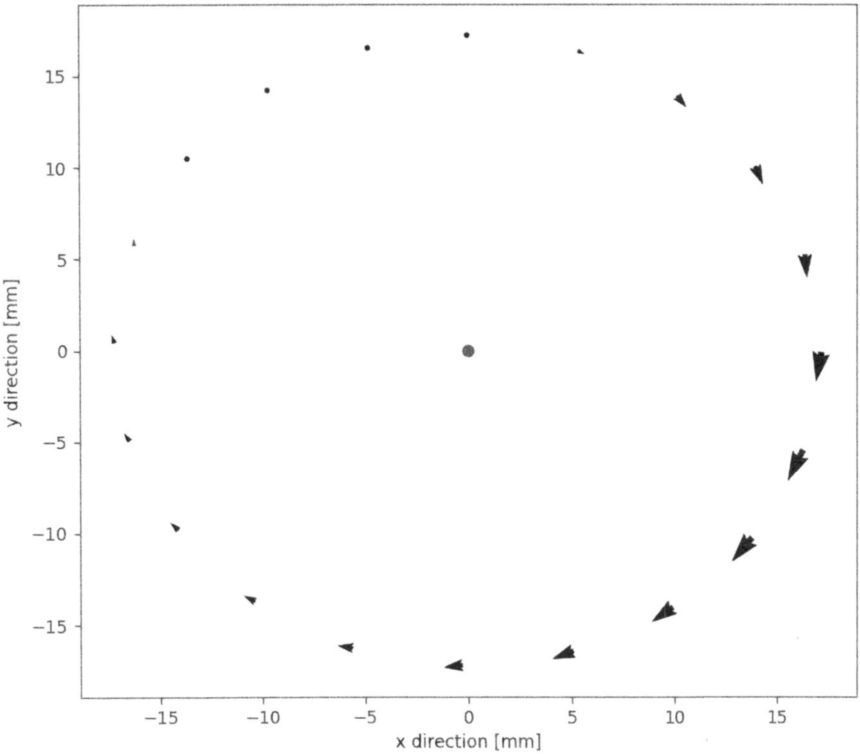

Fig. 3.16 Vector force pedal plot of the author cycling

now excellent free tools such as the *Python Mechanism library* (Morris 2023) that can be used for this purpose.

The code used for this analysis is published in Appendix 1. Initial parameters include the seat angle, seat height, crank length, thigh length, shank length and ankle to pedal axis distance. Additional required (non-anthropometric) parameters include both the pedal speed (rpm) and the pedal/crank force measure from 0° to 360° as shown on the biological clock (Fig. 4.9, p. 85).

Whilst the motion created by any ideal mechanical model does not contain all the subtitles of motion (especially when considering the coronal plane), it is worthwhile presenting constraint equations and the mechanical model for readers who wish to understand and use/modify the code.

The five-bar model is constrained to a 2D plane, and has two degrees of freedom (DOF) requiring two inputs (a synchronized crank and ankle angle) to orientate the mechanical links of the: foot, shank and thigh segment. The five-bar model shown in Fig. 3.18 is of the same construction as described by Hull and Jorge (1985) and Gonzalez and Hull (1989). Arranging the links as shown in Fig. 3.18, intuitively, it can be seen that the two DOF can be defined by two loops for the python

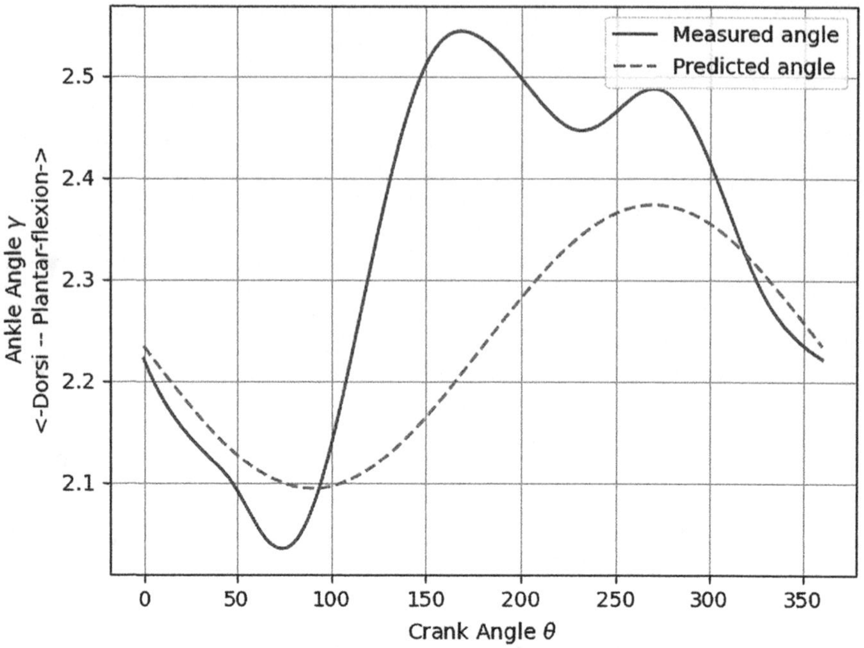

Fig. 3.17 The difference between a measured and predicted ankle angle

programme. The larger loop 1 (Eq. 3.3) describes kinematical summing of the ridged links vectors around the entire chain, whose sign is identified by the direction of summation to the knee joint. Loop 2 (Eq. 3.4) is equal to the sum of the links around the shank.

The Python Mechanism library takes each of the vectors defined in the start position and will advance their position based on a chain of coordinates, calculated from the defined rate (rpm) of crank rotation. Angles are measured from the horizontal which, for the hip angle ($\varnothing_2$), is reasonably typical of what is found in the literature. The knee angle is measured with respect to the thigh as shown in Eq. 3.5. The ankle angle ($\varnothing_3$) may be set at 90°, although there are two options. The ankle angle may be approximated by that of a 5 degrees dorsiflexion, then plantarflexion in the shape of a sine wave (Carpes et al. 2014). The second option is to measure your ankle angle using the *Kinovea* software which will constrain the model to resolve your knee and hip angle when cycling with good accuracy (Fig. 3.17). Hence, the distance from the pedal axle to knee along the dashed line is constrained using the angle $\varnothing_3$ for each time frame, the constraint equation for the ankle knee angle is shown in Eq. 3.6. The code is commented throughout for readers wishing to modify the code and investigate their kinematics for bicycle fitting purposes. However, an alternative approach is to use finite element method which can be implemented using (a free student) software Analysis System (ANSYS) through the mechanism toolbox, which provides a more intuitive method.

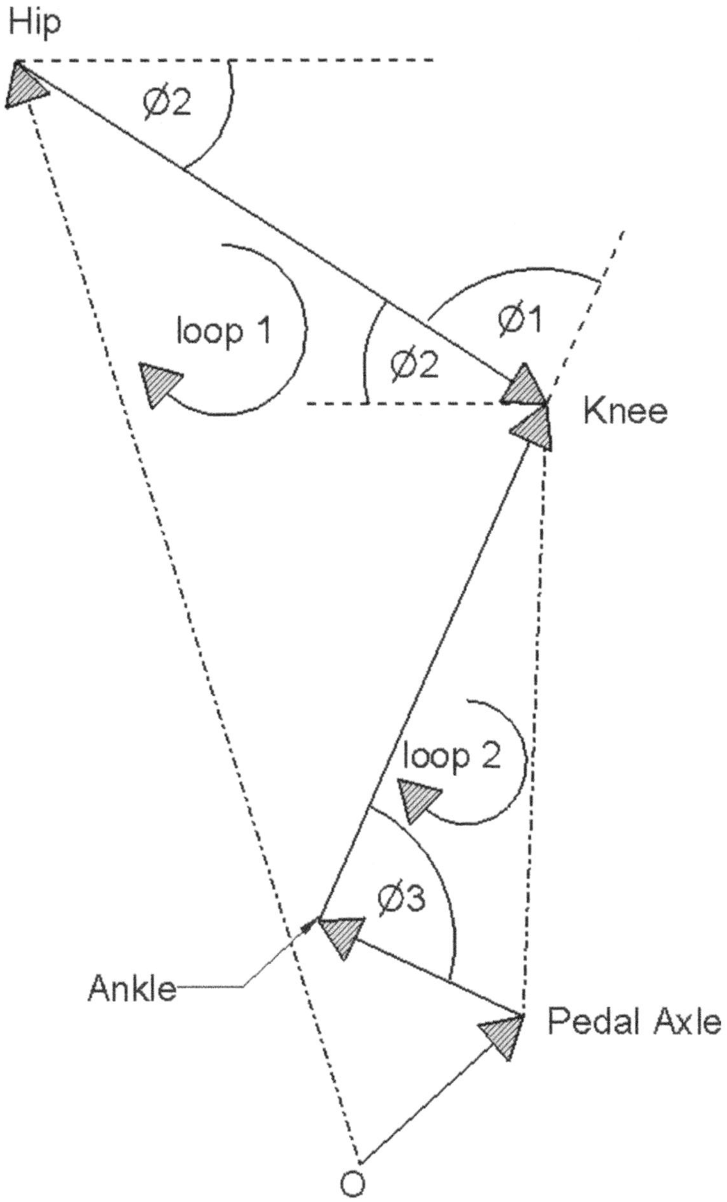

Fig. 3.18 Schematic of the constraints used in the five-bar cycling model

To that end, Figs. 3.18 and 3.19, respectively, illustrate the five-bar kinematic model and angles determined for the thigh, knee and ankle. As expected, the pedal moves in a circular motion, the ankle in an ellipse and knee in an ark. This ark motion illustrates that the shear loads are constantly applied to the knee and

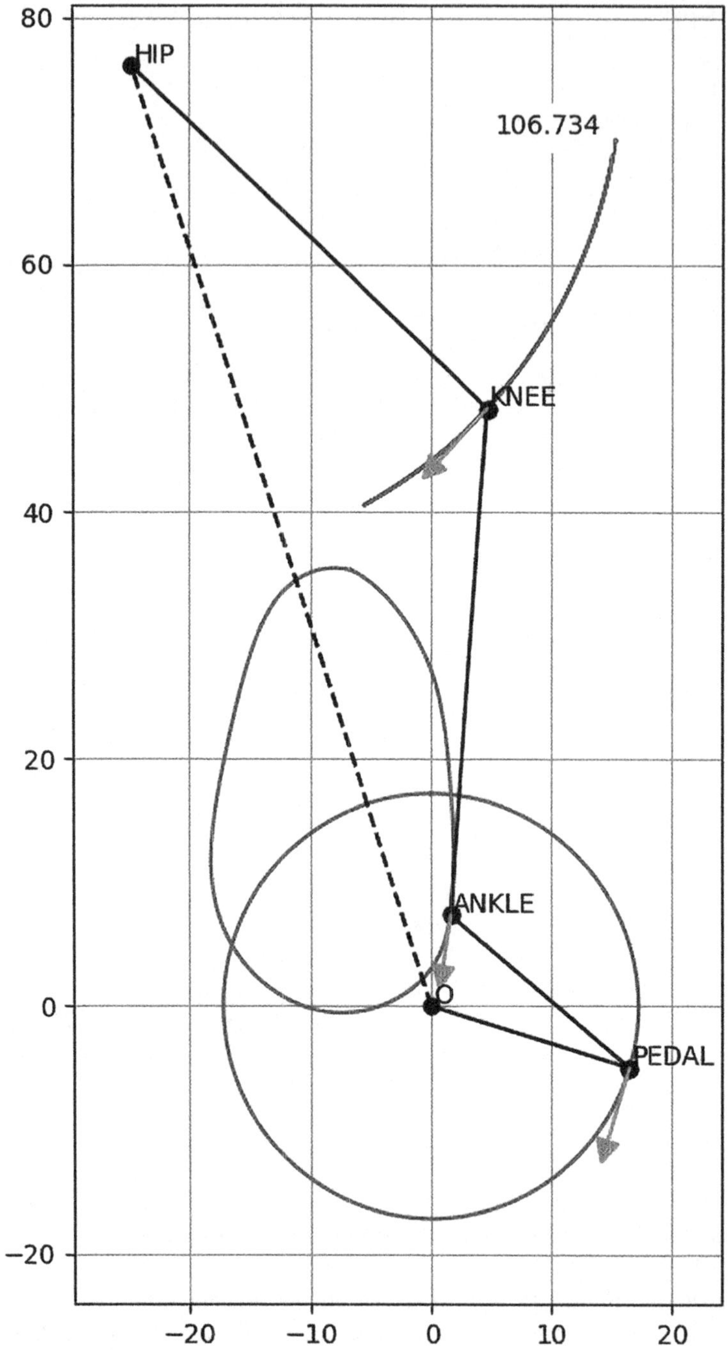

Fig. 3.19 Two-dimensional kinematic model of a single lower limb during pedalling using the ankle angle shown in Fig. 3.17 (p. 60)

highlights why muscle tone is so important when agonistic muscles are working because muscles pulling the joint from one side will cause excessive forces and therefore joint loading.

$$\overrightarrow{R_{HK}} - \overrightarrow{R_{AK}} - \overrightarrow{R_{PA}} - \overrightarrow{R_{OP}} + \overrightarrow{R_{OH}} = 0 \quad (3.3)$$

$$\overrightarrow{R_{PA}} - \overrightarrow{R_{AK}} + \overrightarrow{R_{PK}} = 0 \quad (3.4)$$

$$\varnothing_2 = -\angle\overrightarrow{R_{HK}} \quad (3.5)$$

$$\varnothing_1 + \varnothing_2 + \angle\overrightarrow{R_{AK}} = 180^{\circ} \quad (3.6)$$

3.5.1 *Practical Five-Bar Model Usage*

It is possible to explore the effect that saddle height has on knee flexion although it is assumed that the seat post angle, a line drawn from the bottom bracket centre to the greater trochanter, is approximately the same, and in this case, it is 72°. It is likely this angle will change when positioning yourself on the saddle, for example, when making subtle adjustments on a static bicycle during the pedal revolution. However, such models are useful in that the hip to bottom bracket height can be adjusted until a minimum knee flexion angle of 30° can be observed, indicating an approximate saddle height. Optimal knee flexion angle will be discussed in greater detail in the next section.

Another interesting aspect of mechanical modelling is being able to "play" with a joint angle, for example, the ankle angle during the pedal revolution and its effect on the maximum ankle power. If an ankle angle of 140° is maintained (or fixed), there is no phase shift of the peak's powers generated around the ankle joint compared to the knee and hip, showing the importance of the timing of ankle motion (Fig. 3.20); in comparison with Fig. 3.15, it highlights this difference. There may not be significant ankle motion when cycling, but this model does demonstrate the role of the ankle muscles in delivering power to the pedals with optimal timing during the pedal stroke.

3.6 Bicycle Fit

The practical application of having a fuller understanding of bicycle biomechanics is to become more aware of how you are riding and help find the right compromise between comfort, power production and aerodynamics. The importance that a rider attributes to these variables depends on the type of riding that they are participating in, whether touring, road racing, off-road riding or time trialling. The variables

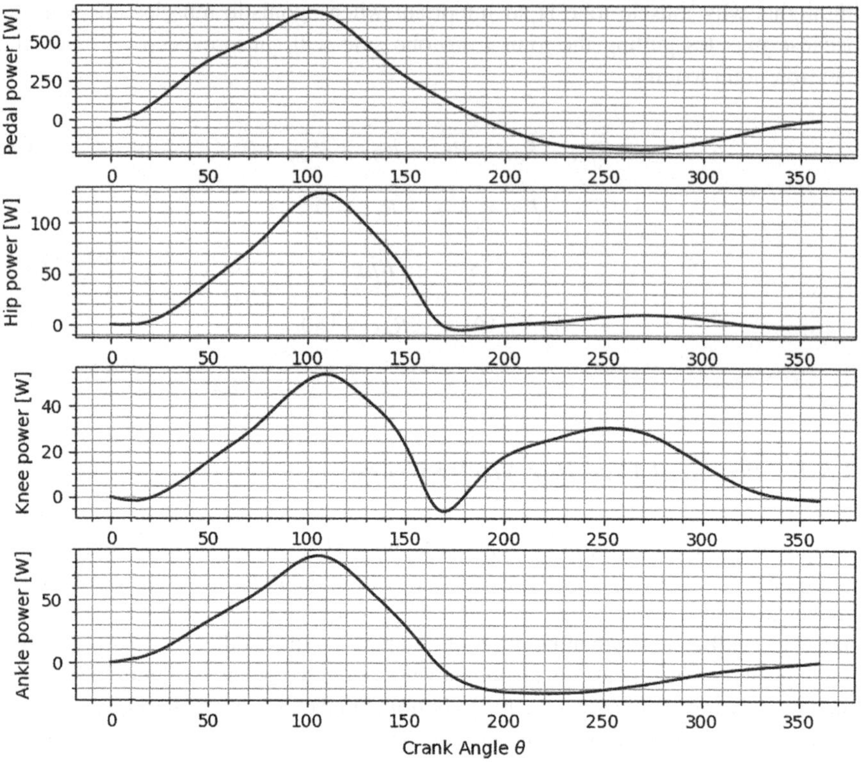

Fig. 3.20 The peak power of the ankle is in phase with the hip and knee power when the ankle is prevented from flexing

include adjustment at the three points of contact pertaining to the bicycle: the handle bars, saddle and pedals. The variables if adjusted to extremes can result in overuse injuries within themselves; for example, a saddle which is set too high to improve aerodynamics may result in Achilles tendinopathy as a result of too much stretch (Silberman et al. 2005). There are many guidelines with respect to fitting the rider to the bicycle and can be broadly divided into static and dynamic procedures as summarised in Table 3.3. However, there are now online dynamic bicycling fitting tools such as *MyVeloFit* which, will be reviewed to check the adjustments provided in Table 3.3.

3.6.1 Saddle Fore-Aft Position

The fore-aft setting of the saddle to a position further forward also reduces the effective saddle height. Thus, both the height and the forward fore-aft position should be considered together. In general, it is advised that when the pedal is in

Table 3.3 Static and dynamic procedures of bicycle adjustment

	Static	Dynamic
Saddle fore-aft position	It is linked to the saddle height and should be set so the ischial tuberosities of the pelvis are supported and the elbows are not too extended too fully	
Saddle height	It is based on a statistical length such as % of trochanter height or ischial tuberosity—another measure is the Lemond method of 88.3% of the inseam of the inside leg (Silberman et al. 2005)	The use of knee flexion of 30–40°
Saddle tilt	Depending on terrain, tilting the saddle can help reduce backpain and improve metabolic efficiency cycling uphill	
Stem height	It effects weight distribution of bicycle and aerodynamic positioning and should be set accordingly with set reach to prevent excessive trunk flexion and/or excessive elbow extension	
Q-factor (inter-pedal width)	It is usually advised that you inter pedal width should match your natural gait width. There is evidence that varying the Q factor does not affect knee extension moments but does in the coronal plane which can indicate susceptibility to injury (Thorsen et al. 2020). The Q-factor of chainset is fixed; thus, cleat positioning gives more opportunity to adjust step-width	
Cleat position	It is generally recommended that the cleat be positioned so the first metatarsal head lies over the pedal axle (Silberman et al. 2005)	

the 90° position, a plumb line drop from the knee patella should pass through the pedal axis which is quite a tight tolerance (Kolin and De la Rosa 1979; Silberman et al. 2005; Bini et al. 2014a). It is the authors' opinion that for a given level fatigue, terrain and wind resistance, it is likely that we will naturally adjust posture and position on the saddle without thinking, to maximise comfort and to maximise muscle usage.

By setting the saddle in a more forward position, it was found that more work was performed at the knee joint, along with reduced work at the hip. There was little evidence that the magnitude of the pedal forces was effective (Bini et al. 2014b). This indicates that there is a fore-aft saddle position which does affect muscle power development, and finding an (average) optimal position that allows adjustment during a ride is required. Moreover, when sitting on a saddle, we balance our weight through our ischial tuberosities, and if not correctly positioned, this will cause a great deal of discomfort.

Therefore, by considering the knee joint loading, using a plumb line from the knee to the pedal should only be seen as a guide, for the reason that all the joints work together to provide the pedal force. In the author's opinion, it would be beneficial to view the direction of pedal reaction force projected from the pedal past the ankle, knee and hip joint in real time by a laser line which would highlight the pull of the joint muscles during pedal cycle. For the reason that the line would move from behind the hip and knee joint at TDC to the front of the ankle joint at BDC, providing an easy illustrative vector will help make adjustments when setting up the bicycle (Fig. 3.21).

Fig. 3.21 Setting the saddle fore-aft position. Plumb line from Patella passes through pedal axis

3.6.2 Stem Position (Height and Length)

The position of the stem ultimately effects a number of variables which include bicycle weight distribution and aerodynamic drag. The biomechanics are affected because the trunk angle affects the maximum hip flexion angle, and therefore the knee and ankle flexion angle. However, reach and elbow flexion is also affected, contrary to the popular belief that cycling in a more upright "tourist position" does not help comfort, and the author has never found this either (Carahalios 2015). The further forward that you lean, there is less weight applied to your ischial tuberosities and more on your arms. There is obviously a balance, but an all-upright sitting position reduces power and comfort.

If you were to stand on a weighing scale and lean forward adopting a cycling position and try and push down with one leg compared to leaning forward and gripping handlebars, say, attached to the wall in front of you, it would be possible to produce more downward force with the latter set-up (Kolin and De la Rosa 1979). Additionally, Kolin and De la Rosa (1979) explained that trunk flexion increases gluteus maximus activity and trunk flexion adopted when cycling, therefore,

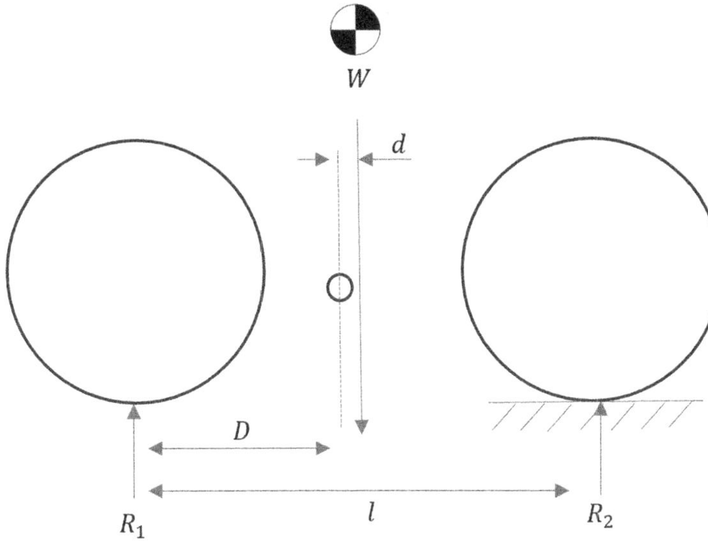

Fig. 3.22 Weighing your bicycle

increases power output. Indeed, while this experiment was conducted practically by feeling the "rear end" muscle movement, there appears to be truth in the hypothesises. Preece and Alghamdi (2021) has shown when walking the hamstring muscle activity increases with truck flexion as well as the contraction of the agonistic muscle group. Too (1994) found that when comparing the power output of 16 male recreational cyclists, at a trunk angle of between 60 and 90°, the power output was significantly greater than at 120°, with the mean power being greater at 90° than 60°. The muscles are connected in the body and because of the leverage effect of the arm the truck muscles, due to handle bar positioning, their optimal output is affected by arm reach.

It was generally found that the rear/front wheel weight distribution for amateur cyclists self-adjusting their bicycles is 61.5/35.5% and 59.6/40.4% for professionals (Carahalios 2015). However, the interesting point to note is that the combined weight of the rider and bicycle generally lies 1 mm ahead of the bottom bracket centre. This aspect of weight distribution is also affected by the fore-aft saddle position, but it is interesting how the rider positions their weight over the turning axis of the cranks—assisting leverage during downstroke but hindering upstroke. Therefore, this may be indicative of a compromise between using body weight to help the downwards thrust but also assist muscle relaxation on the upstroke. The difference in set-up between the amateur and professional rider could also be the result of the perceived difference between comfort and the compromise between aerodynamic drag. However, it does suggest that after the saddle fore-aft position is fixed, setting a wheel on a weighting scale as shown in Fig. 3.22, and adjusting the handlebar reach until the weight acts through the bottom bracket may be an appropriate metric to use to assist bicycle fit.

$$d = \frac{(l - D)R_2 - DR_1}{W} \tag{3.7}$$

In general, for a bicycle frame which fits the rider (not one which is positioned for optimal aerodynamic gain), the stem length will approximately be 10 cm in length. The handlebars will be approximately positioned 2.5–5 cm below the height of the saddle with the elbows flexed to approximately 60–70 degrees in the drop position (Silberman et al. 2005).

3.6.3 Application Fitting Tools

To summarise, rather than directing the reader to Kinovea software to carry out their own kinematic analysis, I wanted to trial a tool which could be used to dynamically fit the bicycle to the rider. However, even if a tool is used, it is good to understand why bicycle fit is important, even if riding is more interesting than the physiology of human movement! To that end, I used MyVeloFit and was interested in the results because while in my 20 s, I participated in occasional club races and was a reasonably fit rider. I am now 20 years older and have poor flexibility after I was hit by a car. On average, I will probably cover 100 miles a week with commuting and gravel riding—the latter being taken up for the reason that my weekend rides are shorter than they used to be, so going off road locally gives more variety.

It may seem that not going by feel when setting up your bicycle is over-complicating matters. How a bicycle feels, however, when cycling around the block or on a few miles round trip to the shops and back may be very different to how it feels on the road against a head wind or over a prolonged hill when you are tired.

There are actually a number of variables to consider when setting up a bicycle such as cleat position, saddle height, saddle fore-aft position, handle bar reach and handle bar height; essentially, they are the points of contact that can be varied. In turn, these variables effect the articulation of ankle, knee, trunk and elbow joints, thus the maximum and minimum flexion and extension angles. Ensuring that the joints articulate with a given range is indicative that the musculature is contacting and relaxing within the correct range, and that joints and the base of the spine are not experiencing excessive loads.

The best positional fitting is done dynamically and not by using statistical formula to position yourself whilst stationary on the bicycle. It is better to be riding and measured dynamically because static formulae are generally based upon specific sets of anthropometrics, such as a group of athletes. These can contain bias. Additionally, riders need a fit that allows movement to a more upright or aerodynamic position when required without predisposition to injury or discomfort.

This is where MyVeloFit excelled as it considered individual flexibility as a rider and explained how and where to place the cleats on shoes. Additionally, MyVeloFit communicated and provided additional information icons (Fig. 3.23), where to

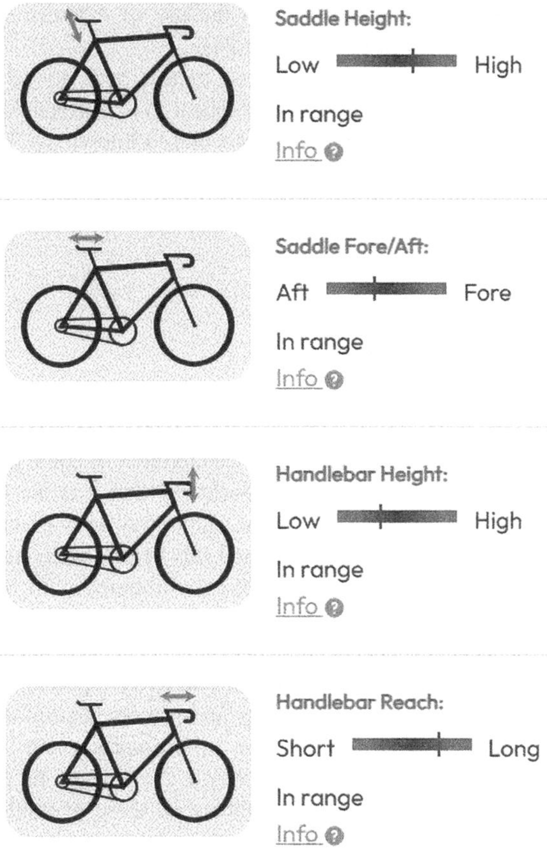

Fig. 3.23 From MyVeloFit—advised changes for positioning saddle and handlebars

Fig. 3.24 MyVeloFit kinematic analysis

position the saddles and handlebars. The ankle, knee, hip, shoulder, elbow and wrist centres were located without the use of markers, and a sagittal plane kinematic analysis was achieved (Fig. 3.24).

Whilst it is not the purpose of this book to validate this tool using objective analysis, the measured angles appear to be correct. Analysis can be completed using a phone camera, and as no specialised knowledge is necessary, this application is thoroughly recommended to fit your bicycle even if you are an experienced rider. There are so many keen cyclists who have poor bicycle fits, and this tool is more than a great place to start!

3.7 Conclusion

As discussed in the proceeding chapters, handling, weight and drag are important, although an optimally positioned engine is of equal importance. While racing speeds are increasing, it is usually portrayed that it is the technology on the bicycle that primarily achieves this. Whilst technology does affect speed by allowing riders to be better fitted to their bicycle, and this better understanding has come from increased knowledge of muscle physiology combined with biomechanics that gives us the tools to bicycle fit based on a number of dynamic variables. It is hoped that reader now has an improved understanding of these variables and where to start looking.

There is no one position that you will adopt on a single ride, and even advancing years will alter your flexibility and perceived comfort; thus, finding a mean bicycle fit will give you room to find a more upright and forward position. Despite there being methods by which today's cyclist can fit themselves dynamically using freeware and power meters (although these methods are not for everyone!); there are now mobile application tools available such as MyVeloFit which can be used to adjust your bicycle correctly.

The disadvantage of biomechanics is that it does not tell you why and how. Why should the saddle height be set to achieve a minimum knee flexion angle of 35°? By studying muscle physiology, the parameters by which we should set the bicycle can be more fully understood. Finally, understanding our respiratory system and how muscles are fuelled and contract to develop force will give the cyclist a greater perception of what limits our muscle physiology and the limitations that our biological system places on muscles to produce power.

Bibliography

Barratt PR, Martin JC, Elmer SJ, Korff T (2016) Effects of pedal speed and crank length on pedaling mechanics during submaximal cycling. Med Sci Sports Exerc 48(4):705–713

Bini R, Hume P, Kilding A (2014a) Optimizing bicycle configuration and cyclists' body position to prevent overuse injury using biomechanical approaches. In: Biomechanics of cycling. Springer, Cham, pp 92–107

Bini RR, Hume PA, Lanferdini FJ, Vaz MA (2014b) Effects of body positions on the saddle on pedalling technique for cyclists and triathletes. Eur J Sport Sci 14(Suppl 1):S413–S420

Carahalios AJ (2015) An analysis of the bicycle-rider interface forces in stationary road cycling. PhD thesis submitted to the faculty of the Graduate School of the University of Colorado

Carpes F, Bini R, Priego Quesada JI (2014) Joint Kinematics. In: Biomechanics of cycling. Springer, Cham, pp 33–42

Elmer SJ, Barratt PR, Korff T, Martin JC (2011) Joint-specific power production during submaximal and maximal cycling. Med Sci Sports Exerc 43(10):1940–1947

Gonzalez H, Hull ML (1989) Multivariable optimization of cycling biomechanics. J Biomech 22(11–12):1151–1161

Hull ML, Jorge M (1985) A method for biomechanical analysis of bicycle pedalling. J Biomech 18(9):631–644

Kolin MJ, De la Rosa DM (1979) The custom bicycle: buying, setting up, and riding the quality bicycle. Rodale Press, Emmaus

Merrill Z, Perera S, Cham R (2019) Predictive regression modeling of body segment parameters using individual-based anthropometric measurements. J Biomech 96:109349

Millour G, Velásquez AT, Domingue F (2023) A literature overview of modern biomechanical-based technologies for bike-fitting professionals and coaches. Int J Sports Sci Coach 18(1): 292–303

Morris G (2023) Mechanism 1.1.7. From https://pypi.org/project/mechanism/

Nordeen-Snyder KS (1977) The effect of bicycle seat height variation upon oxygen consumption and lower limb kinematics. Med Sci Sports 9(2):113–117

Preece SJ, Alghamdi W (2021) The effect of increasing trunk flexion during normal walking. Gait Posture 83:250–255

Silberman MR, Webner D, Collina S, Shiple BJ (2005) Road bicycle fit. Clin J Sport Med 15(4): 271–276

Thornton W, Jackson J (1980) 3 anthropometry and biomechanics. 2024, from https://msis.jsc.nasa.gov/sections/section03.htm

Thorsen T, Strohacker K, Weinhandl JT, Zhang S (2020) Increased Q-factor increases frontal-plane knee joint loading in stationary cycling. J Sport Health Sci 9(3):258–264

Too D (1994) The effect of trunk angle on power production in cycling. Res Q Exerc Sport 65:308–315

Wu G, Siegler S, Allard P, Kirtley C, Leardini A, Rosenbaum D, Whittle M, D'Lima DD, Cristofolini L, Witte H, Schmid O, Stokes I, Standardization and Terminology Committee of the International Society of Biomechanics (2002) ISB recommendation on definitions of joint coordinate system of various joints for the reporting of human joint motion—part I: ankle, hip, and spine. International Society of Biomechanics. J Biomech 35:543–548

Chapter 4
Muscle Physiology and Force Development

4.1 Introduction

Utilising rigid body mechanics to measure physiological outcomes as motion is all well and good, although it is hard to fathom the meaning of joint extension angles to handlebar reach without gaining an understanding of muscle physiology. Muscles have an optimal operating condition of speed and contraction length which defines our physical limits to produce power along with our respiratory system. Therefore, to develop a complete understanding of bicycle fit and the limitations of endurance, the mechanism behind muscle contractions are now explored.

4.2 Muscle Mechanism

A basic illustration of the (striated) muscle found in our limbs is provided in Fig. 4.1; it is unlike the involuntary controlled, smooth, non-striated muscle found in cardiac tissue or blood vessels. The muscle is striated because of the underling structure: composed of cylindrical single-cell muscle fibres that contain sarcomeres. The sarcomeres are arranged in series along the muscle fibre whose structure contains think (myosin) filaments that have heads which bind to thin (actin) filaments; the myosin heads are connected to the (Z-line) wall of the sarcomere by elastic (titin) filaments—assisting in giving the muscle passive elastic properties. It is this repetitive structure that gives the fibre muscle cells their stripy, and therefore striated appearance; the parallel arrangement of muscle fibres ensures that there are trillions of sarcomeres arranged in especially large muscles.

To contract a muscle, we send a signal from a motor neuron: calcium is released to expose the actin-binding sites for myosin heads, a chemical reaction occurs releasing energy in the form for adenosine triphosphate (ATP), and this allows the myosin heads to bind sequentially to the "steps" of the actin filaments, enabling the

A. Crimin, A. McGarry, *Bicycle Biomechanics*,
https://doi.org/10.1007/978-3-031-85712-6_4

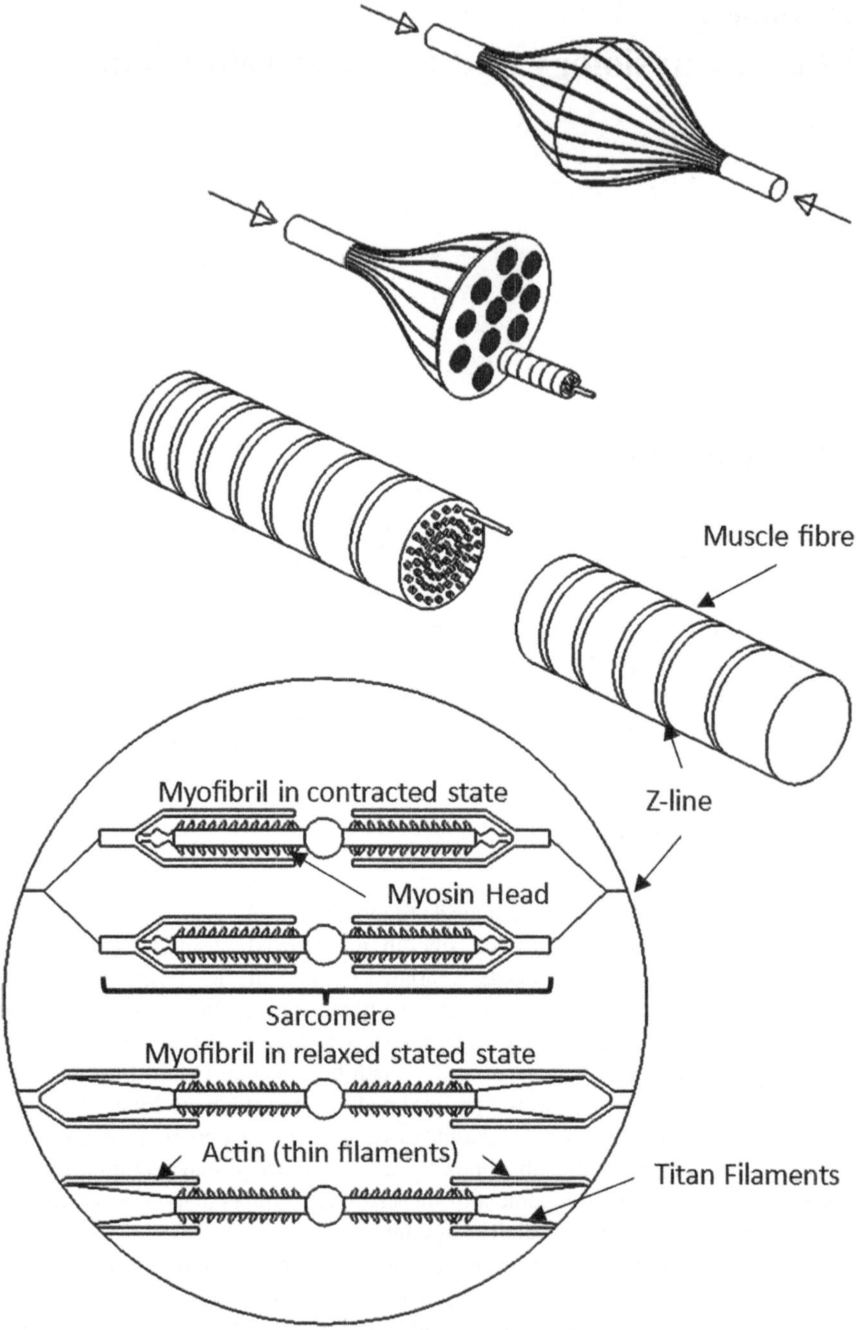

Fig. 4.1 Overview of striated muscles composition

sarcomeres to contract muscles. Hence, when viewing the muscles at the myofibril level making cross bridges to contract the muscle, it raises questions such as how quickly can this process take place and is there an optional working length that the muscle can operate?

Primary properties such as force-velocity, force-time and force-length muscle relationships are well known and are usually evaluated using reductionist techniques; that is, for example, the force-velocity relationship is explored while the force-time and force-length properties are held constant (Martin 2007). The method of changing one variable and keeping all others constant is used throughout science even though it is often found that the variables are all interrelated, and the effect of the measurement may itself change the property of the measured value.

As shown by Martin (2007), the muscle power relationships can be expressed in Equation form (4.1), highlighting the interdependency between variables; that is, the power (P), developed by muscle depends on the length (l), frequency ($1/t$) and velocity (v) of contraction. The frequency of contraction influences the excitation-relaxation kinetics and has a direct impact on the muscle recovery during a cyclic activity. This is of importance to the cyclist because the bicycle can be adjusted to suit the riders' requirements; adjusting the saddle height effects muscle extension length which also effects the pedal frequency, the shorter the crank when pushing the same gear combination will result in a higher cadence. For example, if the same gear combination is selected using two different crank lengths, a 50 tooth chain ring and a 14 tooth cassette gear the shorter crank does result in a overall lower gear thus greater rotational crank velocity.

$$P = F\frac{l}{t} = Fv \tag{4.1}$$

4.2.1 Saddle Height and Muscle Force-Length Relationship

When setting the saddle position, it affects musculature extension and flexion lengths, and when viewing musculature at a microscopic level, it becomes clear why muscles have an optimal working length. The active force developed in muscles is achieved by the sarcomeres, while the (titin) connective proteins in the sarcomeres and surrounding connective tissue play a passive role. Figure 4.2 demonstrates that when the muscle is in mid-extension, the maximal possible area is provided for the myosin heads to bind to the actin sites, thus allowing the muscle to generate its maximal possible force. If the muscle is fully flexed or fully extended, the actin sites for the myosin sites are, respectively, reduced by either becoming blocked or extending beyond reach. Beyond the position of maximal muscle force generation, the passive elastic component of the muscle plays a role in force generation (Fig. 4.3). As it is not practical to view the cyclist muscles in operation at the sarcomeres level, we need to use global measures such as oxygen consumption on a static bicycle in the laboratory to investigate the effect of saddle height.

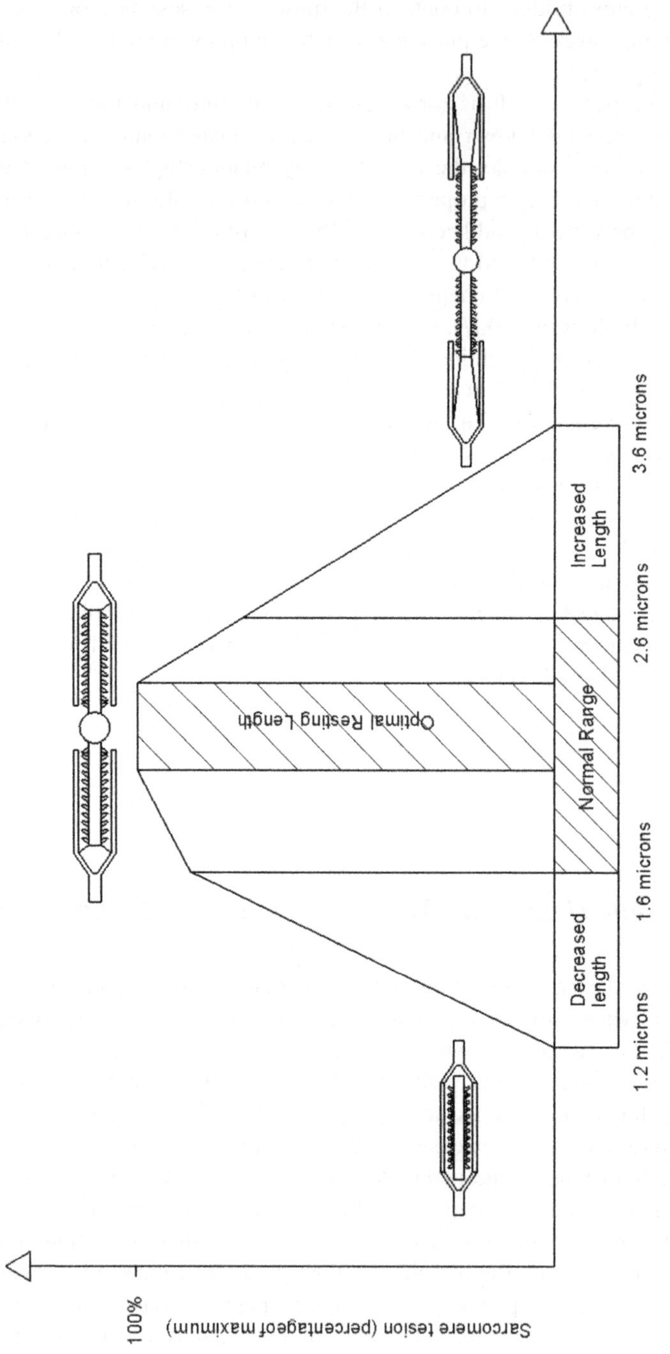

Fig. 4.2 The working length of a sarcomere

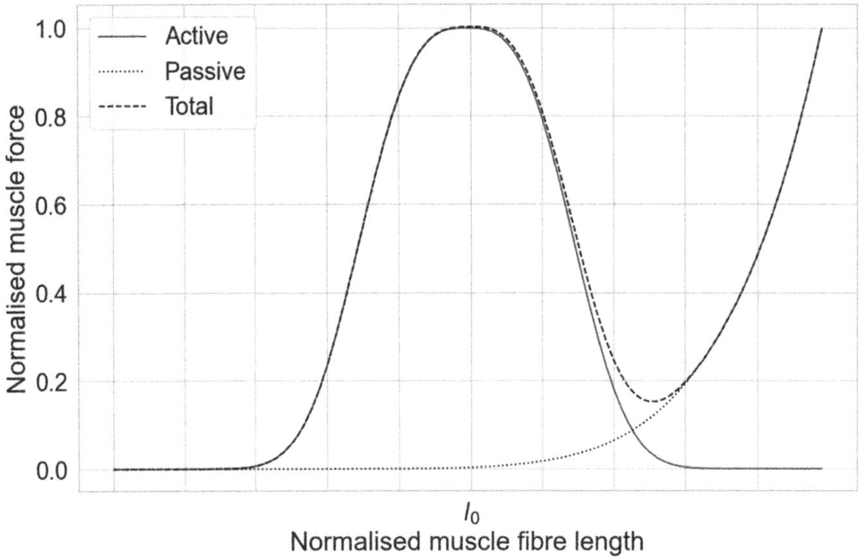

Fig. 4.3 Schematic of the muscle force-length relationship

Shennum and deVries (1976) considered that an optimal saddle height of 103–104% of the pubic-symphysis distance was optimal: it was found that there was a significant maximal increase of oxygen consumption beyond this range. Others such as Nordeen-Snyder (1977), who evaluated female cyclists considered an optimal height of 107% of the pubic-symphysis distance, determined from the position resulting in the lowest oxygen consumption. However, more recent findings such as by Peveler and Green (2011) found that achieving a minimum knee flexion angle of 25° is the most appropriate statistical measure, based on the participant population of 11 well-trained male cyclists. Peveler and Green (2011) evaluated the participants at a saddle set-up height which allowed for 25° and 35° minimum knee flexion, as well as 109% of the height measured from the floor to ischium; the 109% set-up height led to a minimum knee flexion angle falling out of the recommended range (19–44°). Additionally, when Millour et al. (2020) evaluated 17 professional cyclists, the average knee angle fell in with a range of 30–40° for 12 of the participants and 5 with a range greater than 40°. Furthermore, it was found that knee extension angles also mirrored a change of ankle angle variation. More foot plantarflexion was seen with riders who exhibited less knee extension at Bottom Dead Centre (BDC), and more dorsiflexion on riders who would exhibit more knee extension at BDC.

Biomechanical studies which consider a population have a certain bias such as age, individual body anthropometrics, lifestyle, gender and ethnicity to name a few variables. Although other study variables can effect the outcomes such as: population size; is the activity performed on a treadmill, equipment used and era: essentially how do the measurement techniques restrict the participant by effecting their familiarisation period to the surroundings. Hence, when considering that most studies use statistical formula that place the muscle extensions within an optimal operating range of 103–109% ischium height or 30–40° minimum knee flexion when the pedal is at the BDC, it gives the bicyclist a fairly large margin to get the right saddle height. However, getting the wrong saddle height has a significant metabolic cost (Nordeen-Snyder 1977).

From the author's point of view when riding, we constantly shift on the saddle depending on a given fatigue level, wind resistance and terrain. Hence, finding an optimal saddle height is one where the rider can find a position of comfort for their daily riding; it is not just about the power output for the duration of a 1-h study. Fig. 4.4 78 clocks the power output for the author's dominant right during the pedal

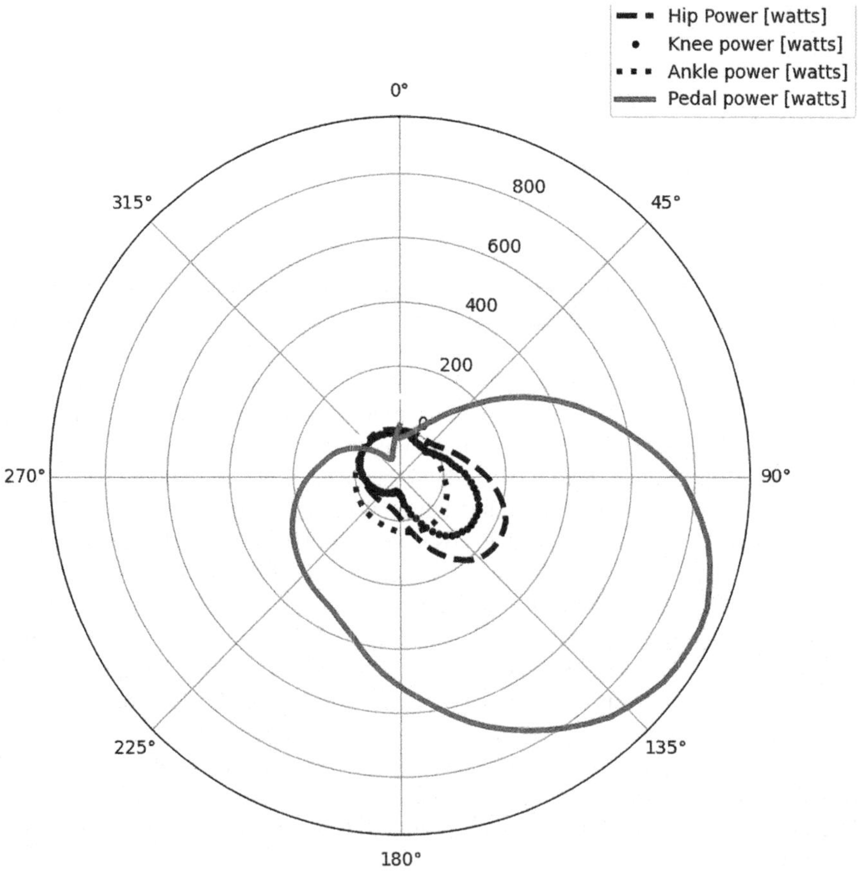

Fig. 4.4 A polar plot illustrating power distribution in watts across different joints during cycling. The plot features four lines: a dashed line for hip power, a dotted line for knee power, a dash-dotted line for ankle power, and a solid line for pedal power. The radial grid ranges from 0 to 800 watts, with angular positions marked at intervals of 45 degrees. The pedal power is the net power transferred down the hip, knee and ankle to the pedal

stroke. It highlights that maximum power is developed after initial leg extension and before the knee is fully extended. This is interesting for the reason that maximum power production does not occur at the extremes: when the leg is at maximum flexion (0°) or maximum extension (180°). Thus, this indicates that it is not the maximum or minimum leg extension that effects power development. In turn, this poses the question if saddle position is also about finding the right position in terms of a compromise between comfort and maximising muscle power output. Indeed, there are other factors such as the muscle contraction velocity and excitation-relaxation kinetics that may also have an effect power production during the pedal cycle. Studies such as by Wilkinson and Kram (2022) have suggested that it is not just the arms that we use to maintain our saddle position, and clearly, we must use the reaction force on the pedals to do so as well. Prior to 2016, the UCI stipulated that for road racing, the saddle had to be set horizontally, and since its relaxation, a maximum allowable saddle tilt of 9° from the horizontal is now allowed. Wilkinson and Kram (2022) evaluated 18 recreational cyclists on an 8° inclined treadmill and also dropped the noise of the saddle by an angle of 8°. It found that there was an average gross metabolic efficiency improvement of 1.4 ± 0.2% measured by evaluating oxygen consumption.

$$\text{Gross Efficiency} = \frac{\text{Work Accomplished}}{\text{Energy Epended}} \tag{4.2}$$

The improvement in metabolic cost was attributed to the fact that when cycling on slopes, effort is used to maintain saddle position, and tilting the saddle helps to reduced that effort; while we do not cycle uphill constantly, it does indicate that there is benefit to titling the saddle down—and the effect saddle position has on other energetics such as maintaining position. Salai et al. (1999) also found that titling the saddle forward by 10–15° helped to reduce backpain, by allowing the spine to maintain a more natural position.

4.2.2 Pedalling Speed and Muscle Force-Velocity Relationship

There are a number of variables that may be adjusted on a bicycle to increase pedal cadence by using short cranks; reducing saddle height or selecting lower gears. Fig. 4.5 shows that the peak muscle force is achieved at a minimum shortening velocity; conversely, at a maximum velocity, the force-generating capabilities of the muscles are reduced. Because the power is a function of the product of force and velocity, (Eq. 4.1), it is possible to plot the muscle power developed during the contraction. It is shown that muscle power development increases to a maximum at 20–30% of the shortening velocity (Martin 2007).

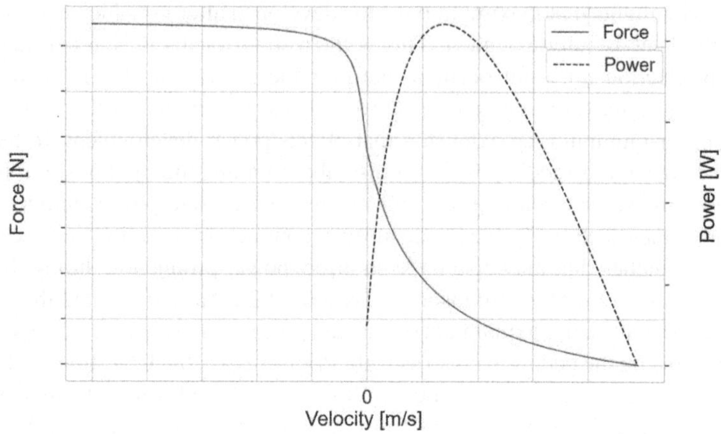

Fig. 4.5 Schematic of muscle velocity-power relationship

As plotted, there is a compromise between speed of contraction and the maximum power that can be developed. What causes this? As the speed of contraction increases, this results in fewer myofibril-making cross bridge connections to pull and contract the muscle (Uchida et al. 2020). Additionally, more energy is also lost as heat due to the increased friction within the muscles. Thus, this is the primary reason why very high pedalling rates of 160 rpm seen by track cyclist are not recommended for the road cyclist (71–92 rpm). Additionally, studies such as those carried out by Hagberg et al. (1981) on professional cyclists showed that they experienced an increased net oxygen consumption and lactate production for pedalling speeds above and below cyclists preferred cadence, and as a result their net efficiency decreased (Eq. 4.3).

$$\text{Net Efficiency} = \frac{\text{Work Accomplished}}{\text{Energy Epended Above That at Rest}} \tag{4.3}$$

This was achieved by treadmill riding and forcing the riders to maintain their effort at two gears above and below their preferred gearing and thus pedalling cadence.

Another consequence of a lower pedal cadence is that the greater force application by the muscle restricts the blood flow, therefore allowing a greater concentration of lactate to build. The muscle pump works by keeping the venous pressure low so that it facilitates return of the blood from the high-pressure arteries. However, even during prolonged exercise, the muscle pump accounts for only a fraction of the flow. Greater muscle forces are thought to reduce this effect (Joyner and Casey 2015). Therefore, it is not surprising for a (short) laboratory study such as Hagberg et al. (1981) that the effect may be more pronounced compared to an all-day event.

4.2.3 Muscle Frequency Relationship

Another substantial limitation related to the velocity of contraction due to the cyclic nature of pedalling is the cadence (Eq. 4.4). For example, at a pedalling cadence of 71–92 rpm, there is only 845–652 ms to get the cranks around one full cycle, which places a substantial time limitation (423–326 ms) on the muscles to develop maximum force on downstroke and recover on the upstroke. It takes 121–400 ms for the quadricep muscle to develop peak tension, and half relaxation time is reported as 76 ms. Indeed, it has been found that at increased crank length, cyclists reached maximum power at lower cadence frequencies. Conversely, maximal power was produced using short cranks at lower pedal speed (Martin 2007). This finding essentially points to the same outcome coming from two directions. If the pedal frequency (f) increases using longer cranks, the crank revolution period (T) must decrease (Eq. 4.4). On the other hand, if the pedal speed increased, the period of the crank revolution decreases so the frequency must increase.

$$T = \frac{1}{f} \tag{4.4}$$

Using this technique of trialling long and shorter cranks, it was found that the results converged to the same crank length in the 170 mm range. Therefore, this indicates the limiting factor of muscle excitation-relaxation kinetics; essentially, the frequencies that muscles can operate at for prolonged periods limit us during endure exercise.

Intuitively, it can be seen that muscle physiology will have an effect on a rider's preference gearing and cadence. Slow twitch dark muscles (that primarily metabolise oxidatively) are more efficient at slower contraction speeds, while fast twitch muscles are less efficient, but have a wider range of working speeds. Additionally, in studies such as Tihanyi et al. (1982), it has also been shown that by studying knee extensor muscles, subjects with a greater percentage of fast twitch muscles can generate more force during contraction. It can be intuitively hypothesised that the proportion muscle types that are employed over a period of activity will depend on factors such as riding speed, fatigue, terrain type and fitness. Many studies do indeed hypothesis this; however, it is difficult to measure the recruitment of fibre types during any physical activity. Other studies such as Umberger et al. (2006) resorted to muscle modelling and have concluded that a greater proportion of fast twitch muscles results in higher cadences. Thus, the general advice is to select a pedal cadence that is natural and comfortable, and not to adopt a cadence out with your preferential pedal speed (Hagberg et al. 1981).

4.2.4 Cycling Musculature

Muscles can only actively shorten during contraction and is the reason why we need an agnostic muscle: that is opposing muscles, for example, such as the biceps to flex the elbow while the triceps are used to extend the elbow, straightening the arm. Similarly, the hamstring muscle group will flex the knee and the quadricep muscle group will extend the knee. If the knee is flexed the hamstring muscles apply tension while the quadriceps are resting; however, they do not go "limp". There is always energy spent to provide muscles tone: that is a tension to enable control so that there will not be sudden jerks, protect the sounding joint by preventing excessive eccentric loading as well as providing shock absorption (Martin 2007). Therefore, while the hamstring muscle group is actively shortening during a (concentric) contraction, the agnostic quadriceps muscles are lengthening during and (eccentric) contraction. Conversely, muscles lengthening under tension are undergoing isotonic contractions. In addition to isotonic contractions there are isomeric contractions where a muscle will usually generate force against an external load and maintain its length. For example, if we extend our arm and place weights on the palm of our hand and maintain our arm position, we are isometrically contracting our bicep muscles to hold that position. Conversely, if our arm extends this is an eccentric contraction.

A pioneering experiment by Abbott et al. (1952) illustrated the difference with startling clarity. The colloquially named Push-Me, Pull-Me experiment was set-up as such that bicycles were fixed back-to-back so that when one rider rode forward, the opposing rider pedalled backwards (Fig. 4.6).

To add to the drama of the experiment a husband (Murdoch Ritchie) and wife (Brenda Bigland) team were employed; the husband being a large strapping Lad who pedalled forward while his small petite wife opposed his motion. It was soon evident that her husband quickly became exhausted and was gasping for breath. Because, the

Fig. 4.6 The push-me, pull-me test set-up described in Abbott et al. (1952)

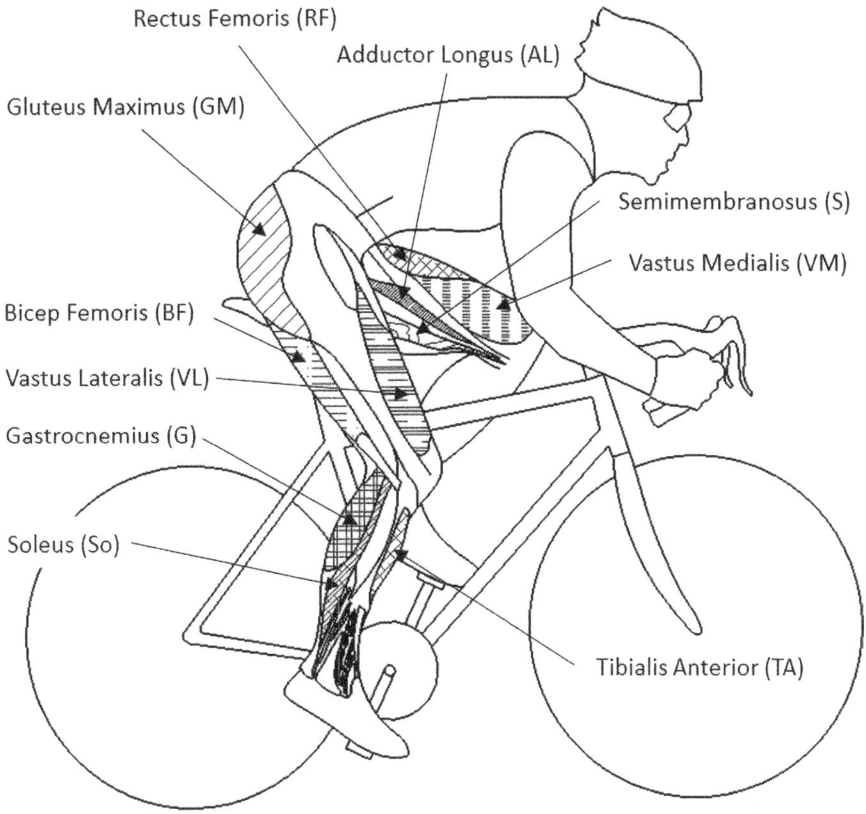

Fig. 4.7 The main muscle in the lower limbs considered when cycling. (Adapted from Burke 1986, p. 76)

muscles of the opposing rider, by pedalling backwards were performing negative work; they did not need to generate power to resist motion as they adsorb power of the rider pedalling forward. In the same manner as described for Fig. 3.14, p. 57, the left limb was relying on the work of the right dominant limb to recover during the pedal cycle recovery phase. To define the type of contraction around a joint the resultant muscle moment (direction of pull) and the joint velocity (direction of movement) should be considered: If the direction of movement and pull are in the same direction positive work is being done, and in opposing direction negative work is done (Gordon et al. 1980).

When we consider the main body of muscles employed during cycling, they are usually the muscles highlighted in Figs. 4.7 and 4.8. The activity of these muscle groups such as the: hamstrings (BF), Quadriceps (RF, VL, VM) or Calf (G, So) muscle groups are readily recognised by the tried cyclist, or after a long ride! Scientists typically identify the patterns of muscle activity using Electromyography (EMG), this technique allows sequence of muscle activation to be measured as well

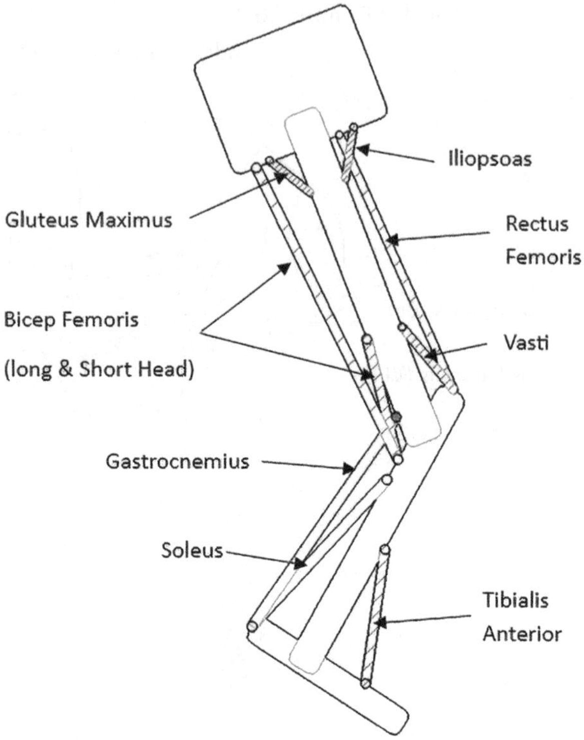

Fig. 4.8 Lower limb muscle attachment points. (Adapted from Papadopoulos 1987)

as the relative magnitude of the muscle activation. The measurements are taken by either sticking electrodes to the skin or using a needle electrode planted into the muscle. Although the latter is often not performed in more recent studies despite measurements containing less noise: for the reason that the technique is painful to implement. The electrode pads only require that the skin is shaved and cleaned and that the electrodes are stuck over the muscle bellies in the most appropriate position. The difficulty of using skin contacting electrodes is that body fat and water content effect the total measurement resistance (thus signal to noise ratio), the other surrounding muscles impulses cause interference. Therefore, is the reason why the main muscles that dwell closest to the skin are most commonly examined (Bini and Carpes 2014b).

The muscle “clock” shown in Fig. 4.9 highlights the main muscle groups which “fire” throughout the pedal cycle, illustrating how the muscle contractions move down the leg from the hip to the ankle during the power phase (0–180°), transferring the power generated distally down the hip to the ankle. It is the role of the biarticular muscles (muscles crossing two joints) such as the hip extensors (biceps femoris), knee extensors (rectus femoris) and knee flexors (gastrocnemius) to transfer power across the joints. Because unlike the muscles that span a single joint, if the hip is extending, for example, the biceps femoris muscle by connection to the hip and also crossing the knee joint will work agnostically with the rectus femoris. Hence, doing

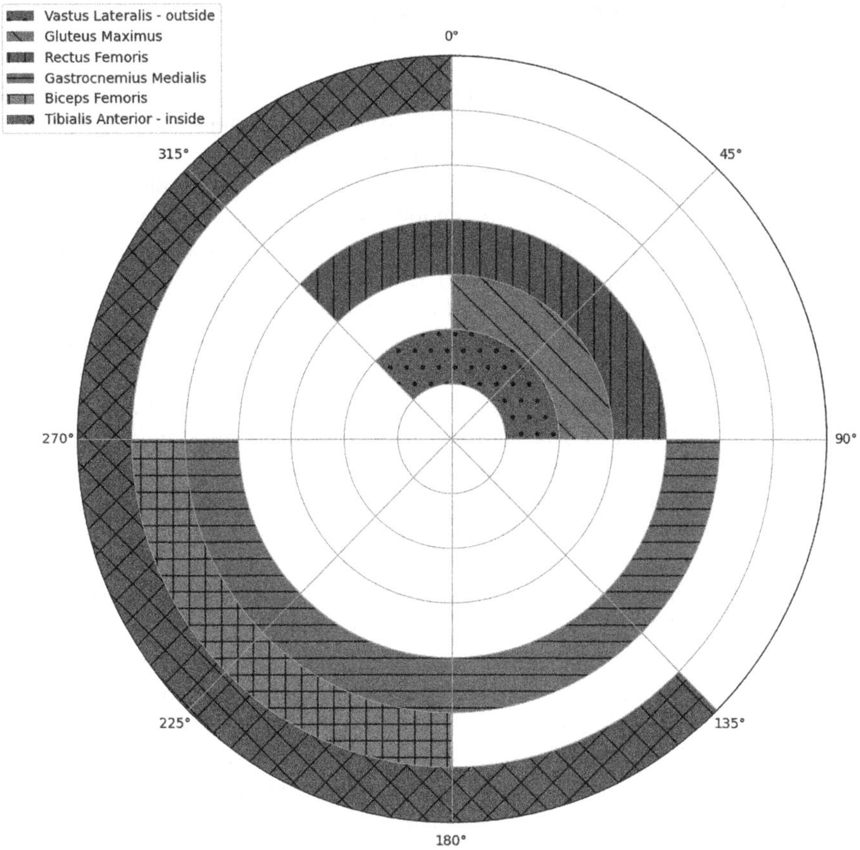

Fig. 4.9 Phase of muscle activation clocked when cycling

so will cause motion of the lower leg (shank), thus assisting the transfer of power down the limb.

It can be observed that the order of muscle employment moves from 0° to 90° where the knee extensors and the hip extensors generate power; then at 90–180°, the ankle plantarflexors transfer power from the hip across the knee to the ankle, and finally at 180–360°, the knee flexors and ankle dorsi-flexors articulate the hip, knee and ankle to move back into position for the power phase (Bini and Carpes 2014a, b). However, analysing muscle activity by EMG can tell us generally the muscles that are used to cause the motion and, to some extent, tell you how hard they are firing; it is difficult to use this type of analysis readily to analyse your performance in terms of asymmetries between the left and right leg, how the forces are being applied to the pedals and quantifying the magnitudes of power generated at the joints. However, to analyse the outcome of all the muscle motion objectively, this is where the kinematics and kinetics of motion using biomechanical techniques become essential.

4.3 Physiology of Cycling and Exercise

There are other aspects of your physiology which cannot be maximised by bicycle set-up and studying biomechanics, include your muscle composition, aerobic and anaerobic capacity and are worth a brief mention.

The form of energy that our muscles utilise is adenosine triphosphate (ATP); the demands of contacting muscles are so enormous that the supply is exhausted within approximately 15 s. When the myosin heads of the muscle break down ATP, they produce adenosine diphosphate (ADP) which can be re-changed to ATP. The infamous chicken legs are always used to explain the difference between fast and slow twitch muscles. The white fast twitch muscles can contact in 0.01 s or less using ATP in the absence of oxygen, producing far more force than slow twitch muscles. The darker slow twitch muscles work optimally under aerobic conditions; they take up to three times longer to generate force and will be predominantly utilised during easy-to-moderate exercise. Due to the nature of their slower contractile speed and good oxygen supply, they can produce ATP on demand. Slow twitch muscle primarily draws on energy reserves from fatty acids and glucose. Conversely, fast twitch muscles draw form energy reserves of glycogen (a type of muscle starch) in the muscles, producing lactic acid as a consequence (Martini and Nath 2010).

Numerous studies have been carried out to establish the main proportion of fibre types of professional cyclists to investigate whether their predisposition for sprinting or endurance is genetically based. It has been found that musculature with a higher percentage of fast twitch fibres work over a greater range of contraction velocities, though less efficiently and in a narrower operating range than slow twitch muscles (Plotkin et al. 2021). It can be imagined that when cycling, the employment of these different types of muscles is dependent on the proportion of the different fibre types, fatigue levels and the type of exertion. Furthermore, there is evidence that fibres have ability to shift between types and that slow twitch fibres can convert to fast oxidative glycolytic (FOG) fibre that is more resistant to fatigue than a fast twitch fibre and can produce more power than a slow twitch fibre (Plotkin et al. 2021). Hence, genetics will play their role in muscle fibre composition although training can influence the percentage composition.

A fundamental aspect of riders' physiology that is always estimated and often cited as a limiting factor of an athlete's performance is their $V0_2$ Max. Essentially, it is a measure of how much oxygen your body can process, and conversely, it also indicates that the limit to muscle power production is not a result of how much oxygen we can get into our body. Typical values of $V0_2$ Max quoted for trained cyclists are 62.6 ± 7.30 mL·kg^{-1} min^{-1} and untrained cyclists 42.6 ± 7.80 mL·kg^{-1} min^{-1}(Hopker et al. 2008); however, cyclists such as Greg Lemond have been reported to have a $V0_2$ Max of 92.5 ± 7.80 mL·kg^{-1} min^{-1}. The best estimate of $V0_2$ Max is to measure the inhaled and exhaled air until exhaustion; there are other statical measures which can be used, such as heart rate to estimate $V0_2$ Max. The $V0_2$ Max can be improved by 20–25% upon exercise and is indeed a measure at how fast you are likely to cycle. However, the cyclist with the best $V0_2$ Max does not always win (Burke 1986).

4.4 Conclusion

The purpose of this chapter was to primarily concentrate on the aspects of muscle physiology that cannot be investigated by the use of biomechanics. By gaining an understanding of muscle physiology, it is possible to gain an insight into how to set-up your bicycle using biomechanics to investigate your pedal style and position.

This has been achieved by presenting the mechanism behind muscle contractions and using published studies that use global factors such as oxygen consumption with measured kinematics to investigate positional factors effecting performance. The field of mechanics has advantages in that muscle powers can be more accurately calculated compared to using EMG, although the muscle physiology helps to explain the mechanism behind the workings.

The concept that resistance to motion and position for power generation should be balanced based on a number of attributes has been developed over the last four chapters. Duration of exercise, age, fitness and flexibility all affect the saddle and handle bar position, and athletes today should have their bicycle fitted based on the riding activity that they are perusing.

The final chapter concentrates on building a dynamometer to analyse the pedal stoke rather than using it as a power cadence meter. By being able to analyse the pedal stoke as a function of force/torque, a greater understanding of limb asymmetries and nuances of power development can be observed. The chapter also highlights the difficulties of accurate kinetic measurement by practical demonstration and provides the reader with commercial off-the-shelf options that have improved accuracy over the main stream dynamometers available.

Bibliography

Abbott BC, Bigland B, Ritchie JM (1952) The physiological cost of negative work. J Physiol 117(3):380–390

Bini R, Carpes F (2014a) Introduction to biomechanical analysis for performance enhancement and injury prevention. In: Biomechanics of cycling. Springer, Cham, pp 1–11

Bini R, Carpes F (2014b) Muscle activity. In: Biomechanics of cycling. Springer, Cham, pp 23–31

Burke E (1986) Science of cycling. Human Kinetics Publishers, Champaign

Gordon D, Robertson E, Winter DA (1980) Mechanical energy generation, absorption and transfer amongst segments during walking. J Biomech 13(10):845–854

Hagberg JM, Mullin JP, Giese MD, Spitznagel E (1981) Effect of pedaling rate on submaximal exercise responses of competitive cyclists. J Appl Physiol Respir Environ Exerc Physiol 51(2): 447–451

Hopker J, Coleman D, Wiles J (2008) Differences in efficiency between trained and recreational cyclists. Appl Physiol Nutr Metab 32:1036–1042

Joyner MJ, Casey DP (2015) Regulation of increased blood flow (hyperemia) to muscles during exercise: a hierarchy of competing physiological needs. Physiol Rev 95(2):549–601

Martin JC (2007) Muscle power: the interaction of cycle frequency and shortening velocity. Exerc Sport Sci Rev 35(2):74–81

Martini FH, Nath JL (2010) Fundamentals of anatomy and physiology. Pearson Education, Limited, Harlow
Millour G, Duc S, Ouvrard T, Segui D, Puel F, Bertucci W (2020) Variability of ankle kinematics in professional cyclists: consequence on saddle height adjustment. J Sci Cycling 9(1):25–32
Nordeen-Snyder KS (1977) The effect of bicycle seat height variation upon oxygen consumption and lower limb kinematics. Med Sci Sports 9(2):113–117
Papadopoulos J (1987) Chapter 2: Forces in bicycle pedaling. In: Rekow R, Thacker VG, Erdman AG (eds) Biomechanics in sport: a 1987 update. American Society of Mechanical Engineers, New York, p 126
Peveler WW, Green JM (2011) Effects of saddle height on economy and anaerobic power in well-trained cyclists. J Strength Cond Res 25(3):629–633
Plotkin DL, Roberts MD, Haun CT, Schoenfeld BJ (2021) Muscle fiber type transitions with exercise training: shifting perspectives. Sports (Basel) 9(9):127
Salai M, Brosh T, Blankstein A, Oran A, Chechik A (1999) Effect of changing the saddle angle on the incidence of low back pain in recreational bicyclists. Br J Sports Med 33(6):398–400
Sarabia Marín JM (2015) Effects of power training with optimal load and repetitions. Doctoral thesis. Departamento de Psicología de la Salud, Universidad Miguel Hernández de Elche
Shennum PL, deVries HA (1976) The effect of saddle height on oxygen consumption during bicycle ergometer work. Med Sci Sports 8(2):119–121
Tihanyi J, Apor P, Fekete G (1982) Force-velocity-power characteristics and fiber composition in human knee extensor muscles. Eur J Appl Physiol Occup Physiol 48(3):331–343
Uchida TK, Delp S, Delp D, Recorded Books I (2020) Biomechanics of movement: the science of sports, robotics and rehabilitation. The MIT Press, Cambridge. 1 online resource (xix, 374 pages): color illustrations, color map
Umberger BR, Gerritsen KG, Martin PE (2006) Muscle fiber type effects on energetically optimal cadences in cycling. J Biomech 39(8):1472–1479
Wilkinson RD, Kram R (2022) Nose-down saddle tilt improves gross efficiency during seated-uphill cycling. Eur J Appl Physiol 122(2):409–414
Wilson DG, Papadopoulos J, Xplore I, Press MIT (2004) Bicycling science. MIT Press, Cambridge

Chapter 5
Building a Power Dynamometer

5.1 Introduction

The purpose of this chapter is threefold: To provide a technical overview of bicycle dynamometers by showing how they work, provide a methodology so the reader has a starting point to build their own dynamometer and present curves which show the kinetics of the power stroke.

Strain gauges are commonly utilised by manufacturers to build bicycle dynamometers as they are a well-understood technology having been around for nearly 100 years. Indeed, there are many other techniques available such as utilising a needle and spring balance on the pedal and using piezoelectric crystals as seen in accelerometers. These techniques may offer improved technical solutions. However, as with all product success, it is not just the technical solutions which make products commercially viable. Other considerations for successful adoption include time to development, design complexities, too much specialist knowledge required, repeatability, ease of manufacture, sourcing parts, profitability and serviceability.

What are the motivations for buying or building your own dynamometer to capture your time series kinetics? Laboratory motion analysis has proven to be invaluable, although it is not the most accessible tool, and it can take a lot of effort to capture good data in an artificial environment, rather than in the outside environment where the action happens. On the other hand, the main advantage that cycling has over a number of human activities, even simply walking, is that it is relatively inexpensive to measure the kinetics of motion.

By empowering the everyday rider to evaluate their cycling kinetics, it also empowers the cycling community as there is a lot of jargon used to sell bicycles which can be filtered by being informed. Additionally, while scientific studies display the general kinematic and kinetic outcomes of pedal stoke and suggest attributes from the captured data that are attributed to cycling efficiency; they usually belong to the well-trained competitive amateur or professional athlete.

A. Crimin, A. McGarry, *Bicycle Biomechanics*,
https://doi.org/10.1007/978-3-031-85712-6_5

To that end, this final chapter begins with a basic introduction to strain gauges so that the reader has a starting point to be able to find more information as required, before presenting the considerations of building a dynamometer by practical demonstration.

5.2 Strain Gauging Techniques

The strain gauge is essentially a variable resistor used to measure the strain on the surface of a substrate; this is achieved by bonding the strain gauge to the surface so the deformation of the part is directly transferred to the strain gauge. This facilitates them to be used in devices such as cranks and pedals for the purpose of force measurement. There are various types of strain gauges: wire, semiconductor and foil gauges, though it is the latter that we are interested in as they are more commonly used in transducer applications and are shown in the examples that follow in this chapter. Foil gauges are composed of a thin metal film bonded onto a thin insulating carrier matrix and are usually thinner than paper; the metal foil is then photo-etched leaving the gauge grid, alignment markers and solder tabs exposed (Fig. 5.1).

If you took a length of wire and stretched it, the diameter would reduce decreasing the outside surface area, and in doing so, it would increase its electrical resistance. However, considering that the gauge used in your cranks, for example, may only cover an area of $<10\ mm^2$ in size; the wire is coiled into many shorter lengths orientated in the direction of the strain, increasing the length, thus sensitivity of the

Fig. 5.1 Strain gauge detail and alignment. (Courtesy of Tech-Measure)

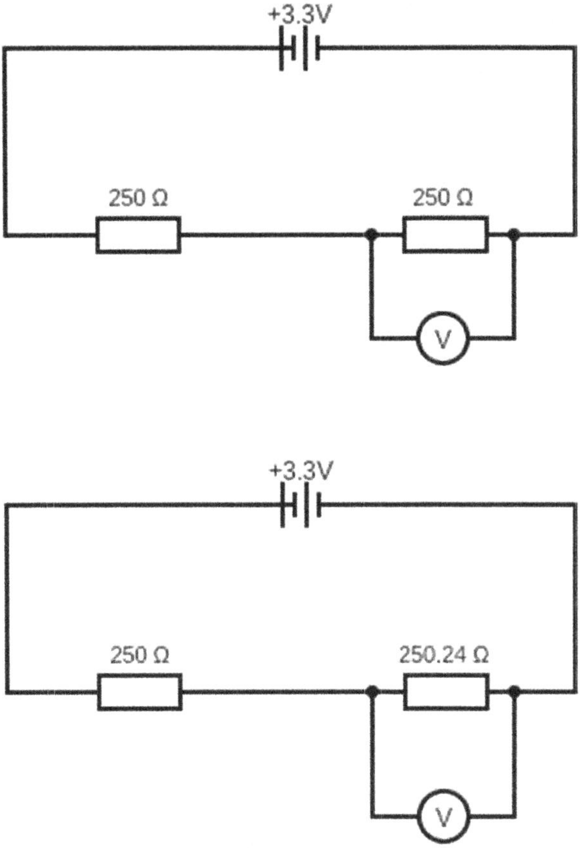

Fig. 5.2 Measuring the resistive change by voltage drop (V)

gauge for the small area being measured. The electrical resistance (R) of the strain gauge can be related to its mechanical attributes of wire length (L), cross-sectional area (A) and electrical resistivity (ρ) by Eq. 5.1.

$$R = \rho \frac{L}{A} \tag{5.1}$$

Hence, it can be observed that the resistance of the gauge increases when it stretches because the cross section of the wire reduces. To measure a voltage change, it needs to be with respect to constant base, that is, electrical ground. For example, if you connect a static resistor in series with a strain gauge with a supply voltage of 3.3 V (Fig. 5.2), there would be an equal 1.65 voltage drop across the resistor and the strain gauge. If the strain gauge experienced a resistance change of 0.24 Ω, the voltage drop across the static resistor and the strain gauge would change (Eqs. 5.2, 5.3, 5.4, 5.5 and 5.6). This is not the most sensitive bridge configuration, although it

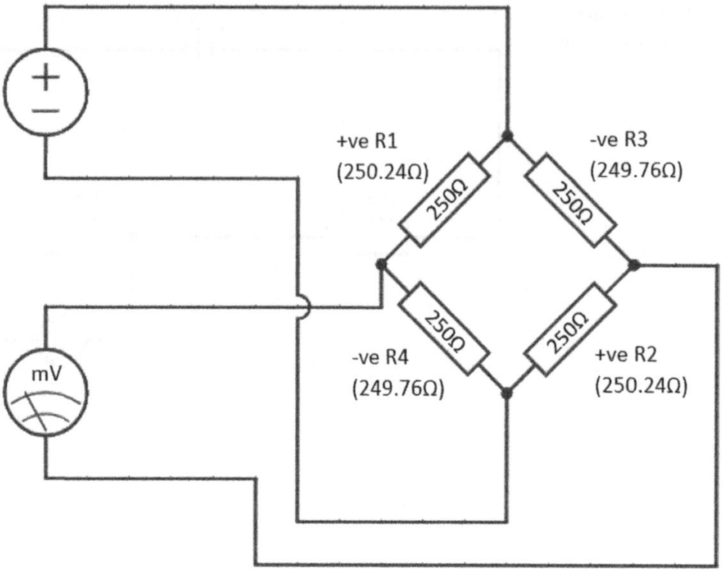

Fig. 5.3 Full Wheatstone bridge arrangement of strain gauges. The resistance changes highlight the direction of the voltage change and where it is measured (V)

gives the idea using realistic values that the observed voltage differences are measured in microvolts.

$$V = IR \tag{5.2}$$

$$I = \frac{3.3}{500.24} = 0.0065968 \text{ A} \tag{5.3}$$

$$V_{\text{strain guage}} = I \times 250.24 = 1.65079 \text{ V} \tag{5.4}$$

$$V_{\text{resistor}} = I \times 250 = 1.64921 \text{ V} \tag{5.5}$$

$$V_{\text{difference}} = V_{\text{strain guage}} - V_{\text{resistor}} = 0.00158 \text{ V} \tag{5.6}$$

Gauges are typically arranged in circuits called quarter, half and full bridges: that is, circuits with one, two and four working gauges, respectively. All these different types of circuits have their own permutations and their own advantages for different types of measurement. However, to this end, the full-bridge circuit will be presented because it is the arrangement that is most likely to be used when building a force sensor; a half-bridge circuit may or could be used, although they are not sensitive. The primary advantages of the full-bridge circuits are bridge balance, temperature compensation and improved sensitivity.

Figure 5.3 illustrates the typical arrangement of strain gauges that you may find in pedal, hub or crank power meters. The positive (+ve) and negative (−ve) signs above the gauge represent how a pair of gauges are positioned, that is, to experience positive (+ve) or negative (−ve) strain. The lettering (R1–R4) on the active circuit corresponds to the positions the gauges would be physically placed on crank arms or pedal spindles as will be shown in this chapter, for example, p.105. The important point to take is that the gauges are referenced against themselves not like that as shown in Fig. 5.2, against a dummy gauge. Therefore, in theory, all the gauges will heat equally once they are powered, with the opposite arms adding and adjacent arms subtracting. Hence, assuming that all of the gauges are subjected to the same temperature, they will all "drift" by the same amount and therefore cancel each other out, leaving the combined mechanical strain as the only output; the initial reading without a load acting is as close to zero as possible. In reality, the gauges all have a tolerance and considering the small voltage differentials being measured, you can appreciate that offsets do occur creating offset in the real signal. However, as shown later for transducer calibration, the strain gauge circuit (microvolt) output differential will be correlated with a unit of measure, so small offsets will be averaged out in the final algorithm.

When the strain gauge circuit is powered, the wires and grid heat up, because they act as a resistance to current; this increase in temperature also causes an additional increase to their resistance which is not uniform across the circuit, causing further signal drift. Signal drift is also exacerbated when working in an outside environment; the ambient air temperature affects the circuit resistance which can lead to a further loss of data accuracy. This is why you should never turn on your power measurement device and start riding; always let your device to warm up and then press the balance/tare zero button.

Another important aspect of having four active gauges when building a sensor is the bridge sensitivity; the voltage difference measured across the full bridge for the example shown in Fig. 5.3 is given by Eq. 5.7 and is 0.003168 V.

$$V_{\text{out}} = \frac{V_{\text{supply}} \delta R}{R} \tag{5.7}$$

This value while greater than the quarter bridge output (0.00158 V), it is still too low to be measured directly by a data logger. For example, Hi-Fi systems without a built-in amplifier cannot oscillate the diaphragm of the speaker. Strain gauge signals need amplification given their very low voltage differentials; hence, it can be appreciated that an amplification of up to 1000 times is usually required, which causes its own difficulties, and the noise is also amplified. Additionally, the offsets caused by the lead wires, strain gauge tolerances, temperature changes, inaccuracies with the data logger and fluctuating battery voltage, to name a few variables, all cause their own problems; and this is before considering the physical aspect of an imperfect alignment. If the geometry is curved and the gauge (as expected) can never be perfectly aligned to measure a load in a known direction, it results in loads from other directions and non-linearities being measured. Thus, you do not have to look very far online to find countless blogs/videos comparing pedals and cranks with 10 percentage differences, and I am not surprised: the aesthetically pleasing geometry of commercial products effects their function. If you are interested in investigating your biomechanics with an off-the-shelf solution, find cranks with minimal features, like they should belong in a laboratory, then you will have a chance that they will actually work with minimal error (p. 55).

Hence, to this end, before detailing how to build a power dynamometer, an overview of where to measure strain, the example circuit used in this book and method of data collection and calibration, will be discussed.

5.3 Data Acquisition

5.3.1 What to Measure

Before positioning the strain gauge circuit, an analysis of the forces acting/boundary conditions should be performed. Generally, commercial solutions measure force input measure at the pedals, cranks and hubs or by "Spider" chain rings. The main difference between hub measurement and other methods is that the force is measured downstream of the chain. Moreover, hub measurement requires more than just modification of components as illustrated in this book and requires component design, manufacture and assembly. However, a novel position using the bottom bracket shaft to measure the riders' input when cycling will be presented and also provides a trace that is more akin to hub measurement.

Figure 5.4 illustrates the boundary conditions that act on the cranks as a result of the pedal force in 2D planar motion.

Indeed, forces do act latterly on the pedals and are of scientific interest; however, they are beyond the scope of study in this publication. The force applied by the foot on the pedal is in a different frame of reference to the effective and radial crank force, that is, the crank rotating frame of reference. Equations 5.8, 5.9, 5.10 and 5.11 show a 2D relationship between the pedal and crank force.

$$\mathrm{F}_{\mathrm{rotated}} = \begin{bmatrix} F_{\mathrm{axial/radial}} \\ F_{\mathrm{effective\ force}} \end{bmatrix} \tag{5.8}$$

$$F_{\mathrm{applied/pedal}} = \begin{bmatrix} F_x \\ F_y \end{bmatrix} \tag{5.9}$$

$$\text{2D rotation matrix} = \begin{bmatrix} \cos\theta & -\sin\theta \\ \sin\theta & \cos\theta \end{bmatrix} \tag{5.10}$$

$$\begin{bmatrix} F_{\mathrm{axial}} \\ F_{\mathrm{effective\ force}} \end{bmatrix} = \begin{bmatrix} \cos\theta & -\sin\theta \\ \sin\theta & \cos\theta \end{bmatrix} \begin{bmatrix} F_x \\ F_y \end{bmatrix} \tag{5.11}$$

The power developed by the cyclist is calculated by taking the product of the moment (M) and crank angular velocity (Eqs. 5.12 and 5.13). The relationship between the number of crank revolutions per minute (rpm) and the angular velocity is shown by Eq. 5.14.

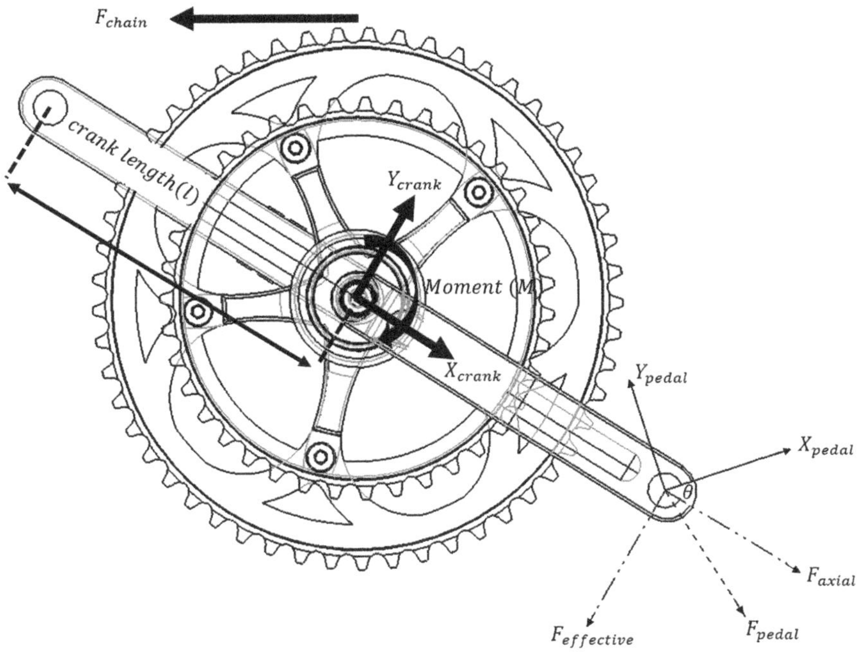

Fig. 5.4 Resolution of forces along the crank frame of reference. The crank system is bold-black, the pedal system solid black and the pedal force dashed

$$M = \text{crank length (l)} \times F_{\text{effective force}} \tag{5.12}$$

$$P = M\omega \tag{5.13}$$

$$\omega = \frac{120\pi}{\text{rpm}} \tag{5.14}$$

Hence, it is shown that pedal measurement requires a method to measure the pedal angle with respect to the crank as in commercial systems (Rowe et al. 1998). However, if every pedal and crank thread started in the same position resulting in the same pedal shaft orientation when fitted to a crank, there would be no need to measure the pedal orientation. However, it will be shown that if building a non-commercial solution, it is possible to negate the requirement to measure pedal orientation with respect to the crank. Thus, by measuring the cyclist pedal force in the crank frame of reference, the effective force is measured directly.

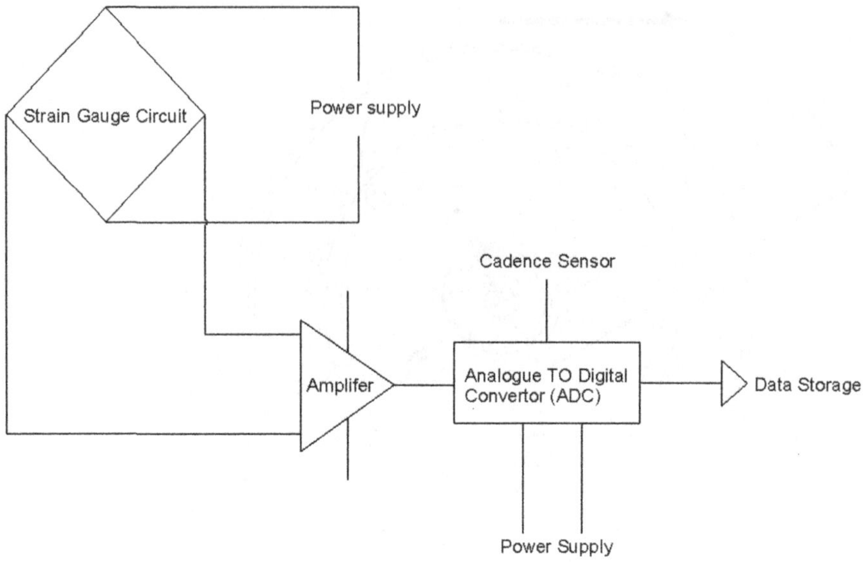

Fig. 5.5 Schematic of dynamometer system

5.3.2 *Circuit Schematic*

The set-up shown in the schematic circuit in Fig. 5.5 highlights the basic building blocks for strain gauge amplification and recording and shows that a cadence measurement is also required. As discussed on p. 93, amplification is required along with an analogue to digital conversion, data acquisition and storage—which can be achieved with integrated "off-the-shelf" *Arduino* hardware components. Additionally, with appropriate component selection, the battery can be connected directly to the board and regulated as shown in the examples in this book with an input voltage of 3.3 volts. However, the bridge voltage is normally chosen with respect to the gauge type and size to help minimize self-heating effects. The amplifier will be powered by the board, and the strain gauge circuit is directly connected to the amplifier.

5.3.3 *Signal Amplification*

The HX711 amplifier is recommended as it provides an economical, small and reliable method of amplification as well as being user-friendly. There are many boards available online although the main criteria to check are that its sampling rate is set, or can be set at 120 Hz, which is appropriate for most biomechanics applications. To digitally capture an analogue signal oversampling by a minimum of five times is required, and ideally it should be ten times. Therefore, capturing at

120 Hz will give reliable rider data at 12 Hz to 24 Hz. Therefore, when cycling at 120 rpm (at a pedal frequency of 2 Hz), the signal is over-sampled by 6 to 12 times of what is required, just giving enough resolution. Many of the HX711 boards available have the following set amplification gains of 32, 64 or 128 of the full-scale differential input voltage—that is, the gain of the voltage difference measured across the strain gauge circuit output. All the signals presented in this book were connected as such to provide a gain of 128 and logging frequency of 120 Hz to provide a "clean" signal.

5.3.4 Data Logging and Hardware

The Arduino *Adafruit Feather M0 Adalogger* was chosen as a battery could be connected directly to the board by a Japan Solderless Terminal (JST) connector and also charged by the board when connected to an external power source without the need of an additional charging hardware. Additionally, data can be saved to a MicroSD card mitigating the need to use telemetry such as Bluetooth. Furthermore, the board can be programmed using an *Arduino* Integrated Development Environment (IDE), which is free and provides inbuilt examples. Therefore, if the reader wishes to add other features such as telemetry, this can be achieved by searching the vast resources of published examples used for Arduino programming. However, saving to the card in the author's experience provides a simple repeatable method and allows for an improved data sampling rate. The *Adafruit Feather M0 Adalogger* is programmed using the *Arduino* IDE; the code shown in Appendix 2, provides an example capturing two strain gauge channels.

To measure the cadence, a reed switch and magnet were used—a switch that closes when a magnet passes (Fig. 5.6). This method was chosen as it is reliable,

Fig. 5.6 Magnet on chainstay used to measure pedal cadence and crank orientation. The Reed switch is set behind the crank

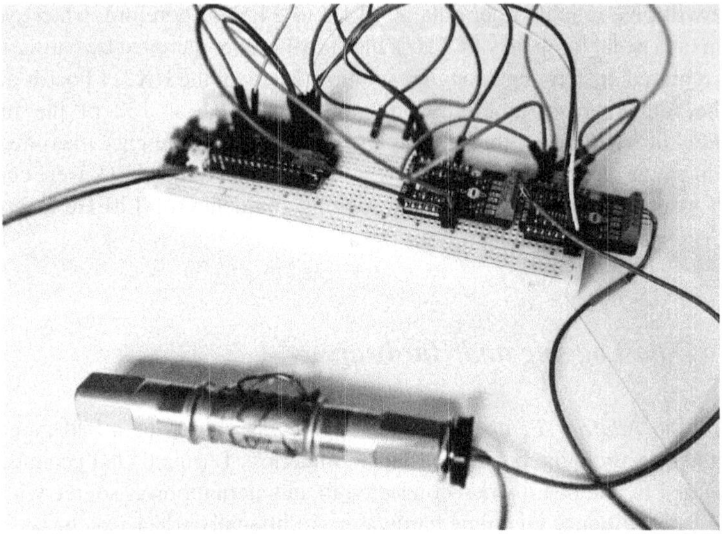

Fig. 5.7 Illustration of the strain gauge circuit (bottom) on shaft, two amplifier boards (for two channels) and data acquisition board (above)

does not consume power compared to other solutions, is the simplest method to install and does not require filtering to obtain clean data.

Figure 5.7 illustrates the breadboard layout using two HX711 amplifiers and the *Adafruit Feather M0 Adalogger,* with strain gauges bonded to a bottom bracket square-tapered shaft. The code (Appendix 2) indicates the pins that were utilised and can be easily changed to accommodate different pin selections on other boards.

The following sections will provide more background into how the gauges should be positioned and wired so that the reader can approach a third party to adhere the strain gauge. Companies such as *Techni Measure* and *QUAD I* in the United Kingdom, respectively, work together to provide strain gauges and fitting service, for both industry and private individuals. Unless the reader has experience with strain gauges, you will find it easier and less expensive to send your chainset or bottom bracket shaft away. Additionally, buying cheap gauges and glue will not give good results, and there is a lot of associated equipment required. However, another invaluable resource is the British Society for Strain Measurement (BSSM); there are a variety of membership levels, and joining this would give access to a variety of resources from asking experts for advice, training courses and literature. Figure 5.8 illustrates how the board and components shown in Fig. 5.7 can be packaged onto the crank by either milling a pocket/slot into the back of the crank and filling with rubber or shrink wrapping the required hardware to the crank. The authors' solution allowed the kinetics of pedalling to be measured in a robust manner and then removed from the bicycle once finished—it was not intended that the crank remain on the bicycle. Table 5.1 summarises the components used to capture the results that will be presented in this book.

Fig. 5.8 Packing boards and strain gauges

Table 5.1 Required hardware excluding a chainset or bottom bracket to build system shown in p. 96

Component	Purpose of hardware
Strain gauge	To measure a strain that can be associated with an input force
Data acquisition board	To read the values from the amplifier
Micro SD card	Hardware to save measured values
HX711 amplifier	Amplify the analogue Wheatstone bridge differential voltage
Magnet and reed switch	To provide a pulse when the crank passes a known position

5.3.5 *Data Collection*

Even before you have packaged and fitted your dynamometer to your bicycle, you may want to test that it can be calibrated. There are many methods of calibration: that is, methods by which you can measure the magnitude of the signal output with a known weight. The method adopted for calibration in this chapter is the dead weight method of applying known static weights and measuring the signal output by either four-point bending, cantilevering and tension/compression. However, the more strain gauge circuits you have, the more elaborate the method of calibration: if you were to start measuring three forces and moments (as is the case for a six-axis load cell), applying a torque, bending moment or shear and only exciting one channel at a time requires planning. The maximum number of circuits presented in this book is two; the gauges are positioned so that they can also be calibrated with minimal crosstalks. For example, gauges that are positioned to measure the bending of the crank in the direction of the pedal force are also influenced by crank twisting and lateral bending. Further discussion of the crosstalk phenomena will be addressed in the examples that will follow in this chapter. Other factors such as linearity and hysteresis in the authors' experience should be insignificant when the bonding of a strain gauge is done correctly to an aluminium metal substrate working at ambient temperatures. In summary, the most important factors to consider when building a strain gauge dynamometer are the following:

1. Strain gauge positioning: Without appropriate positioning, the signal is not measuring what you think it is, and if you cannot use the boundary conditions by correlating the signal with a known force, all further effort is futile.
2. Bonding the gauge: If the gauge is poorly boned, the signal output will be nonsensical.
3. Calibration: It is important to carry out a calibration with the best practices such as considering the accuracy of applied weights and applied direction.

5.3.6 Calibration

To carry out a calibration, it requires applying a force in a known direction—and in the case of strain gauging, the direction in which the gauges are measuring strain. For the cranks presented in this book, it was not possible to measure the crosstalks because, to account for, say, the crank twisting effecting the twisting of the crank should also be measured. Figure 5.9 highlights that the force acting on the end of the crank causes bending in directions that will be picked up by gauges that are meant to only measure the effective force. These additional deflections act to either increase or reduce the measured force and therefore power recorded. Another difficulty

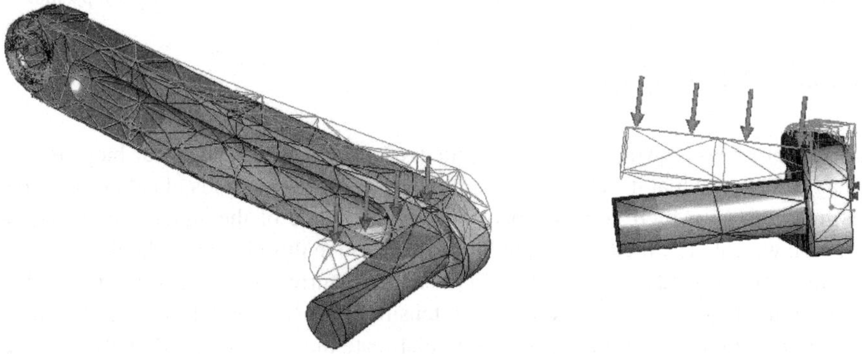

Fig. 5.9 Displacement of a bicycle crank arm during pedal loading. The pedal force induces lateral bending and twisting as well as bending in the direction of rotation

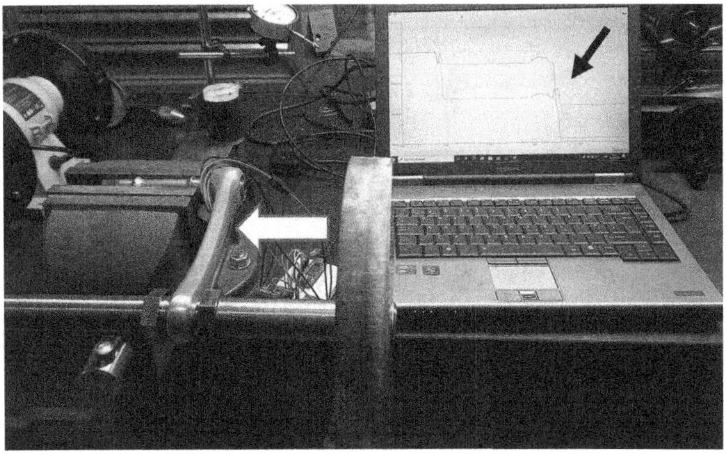

Fig. 5.10 The change in signal output as the weight is slid across calibration bars

Fig. 5.11 Calibration weight positioned next to crank

presented by cranks is that the geometric profiles/styling creates non-uniform strain fields, which create additional non-linearity errors in the measurement. There are certainly ways around this when strain gauging cranks and as shown in McAinsh (2014), as positioning can be optimised to minimise these effects.

The effect of crosstalks acting on the cranks can be visualised below with the following set-up (Figs. 5.10 and 5.11). By using calibration bars that lock on the pedal thread, equal weights can be placed either side of the crank to prevent twist or on a single side and slid to see the effect on the measured signal. Additionally, if a single weight is placed on a single side, then swapped to the same symmetrical position on the other side of the crank, there should be no change in signal. This basic demonstration clearly highlights that crosstalks are present.

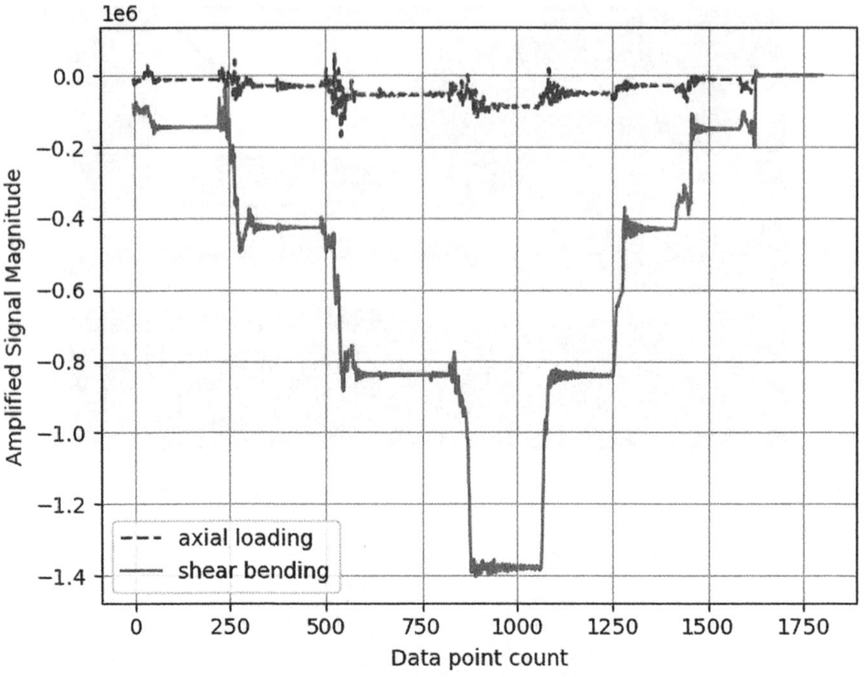

Fig. 5.12 Change in signal output for bending channel as 7, 14, 21 and 28 kg mass added to calibration bars

Figure 5.12 illustrates how the crosstalks can be quantified by, for example, measuring the bending strain as shown in Fig. 5.11 and measuring the radial strain at the same time and vice versa. This indicates that the gauges are aligned appropriately and the calibration set-up is also correctly applying load in the right direction. Moreover, when the applied weights are removed, the signal should follow the same symmetrical steps with no sudden shunting, offsets and equal stepping; this is also indicative that the gauges are well bonded and are not suffering from non-linearity issues.

By noting the amplified voltage difference plateaus and known applied mass, a straight line curve of signal magnitude against the applied force/moment can be plotted. This can be done for the main signal and crosstalk signal, so it is possible to account for inaccuracies created by crosstalks (Fig. 5.13). Indeed, the crosstalk signals can be ignored, and an account of the main channel output (S) is taken only during data acquisition by considering the gradient (C) of the main channel (Eq. 5.15).

$$F = \mathrm{S}/\mathrm{C} \tag{5.15}$$

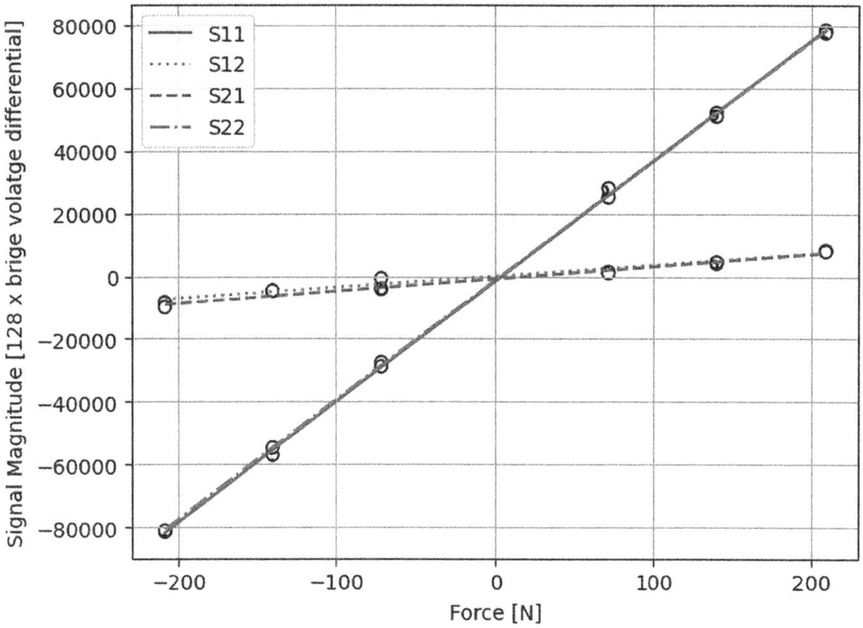

Fig. 5.13 Calibration curves of main signal and cross-talks whose gradients are used for the calibration matrix

However, by measuring crosstalks (C_{21} and C_{12}), it is possible to, as in this case for two channels logged in this example, to express them by linear superposition (Eqs. 5.16 and 5.17)

$$S_1 = C_{11}F_1 + C_{12}F_1 \tag{5.16}$$

$$S_2 = C_{21}F_2 + C_{22}F_2 \tag{5.17}$$

These calibration gradients can be arranged in matrix form as shown in Eq. 5.18, and the inverse of the calibration matrix $[C]^{-1}$ is used as shown to account for the main (leading diagonal) and crosstalk signals.

$$[C] = \begin{bmatrix} C_{11} & C_{12} \\ C_{11} & C_{11} \end{bmatrix} \tag{5.18}$$

$$\begin{bmatrix} F_1 \\ F_2 \end{bmatrix} = [C]^{-1} \begin{bmatrix} S_1 \\ S_2 \end{bmatrix} \tag{5.19}$$

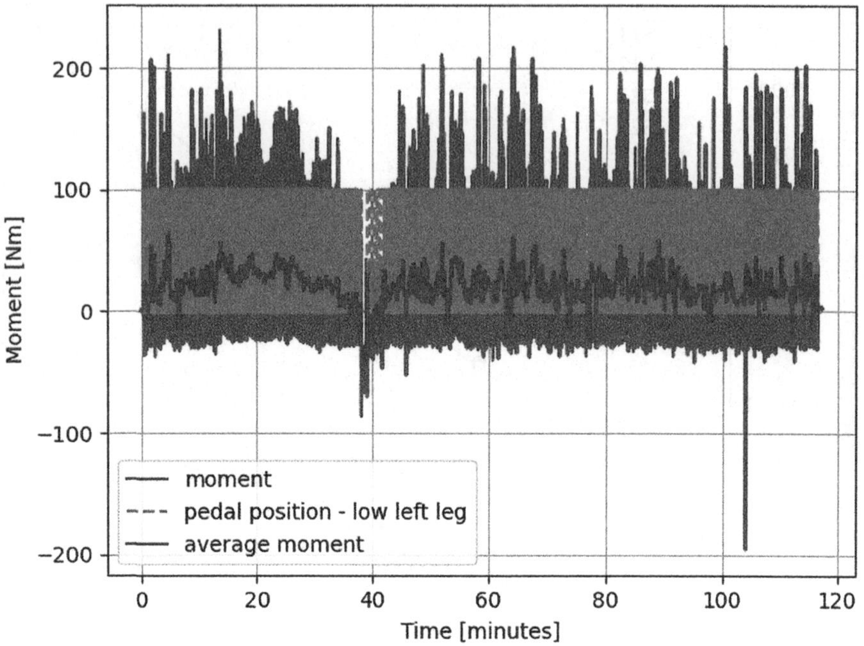

Fig. 5.14 Data collection example

5.3.7 Analysis

Once the dynamometer is calibrated and fitted over, 300,000 data points will be captured for a ride that is under 2 hours, and when plotted, it will look something like the one shown in Fig. 5.14. The easiest method by which to process the data is to use scientific computing (excel will struggle); an example is provided in Appendix 3, using *Python*. Indeed, it may not be necessary to capture the entire ride as the subtitles of the kinetics of a pedalling technique may be identified within a shorter period such as riding up a hill.

5.4 Building a Power Dynamometer

5.4.1 Crank Dynamometer

Crank dynamometers are probably the most intuitive place to start, and on first observation, they certainly have many advantages. The forces of interest can be measured directly with respect to the rotating axes of the crank coordinate system; the orientation of the crank can be measured using a gyroscope, potentiometer,

Fig. 5.15 The combined loading of a bicycle crank arm

encoder or (magnetic) switch, the latter being the simplest method to determine the angular velocity and assumes that the velocity is constant throughout the pedal stoke.

However, it is not intuitively obvious that while a left- and right-hand crank system do measure the force applied by the respective right- and left-hand limb, the left crank on the downstroke also measures reaction torque of the right leg on the upstroke and vice versa—which is what you want because the reaction torque will tell you that the working leg is helping the recovering leg. Second, bicycle cranks are not the easiest surfaces to measure strain for the reason that they bend in directions other than the required direction, and gauges positioned to measure one direction, pick up bending and tension/compressions from other directions: this is known as crosstalk. Figure 5.15 illustrates that the pedal is essentially on a leaver arm and the force does not necessarily align with the crank, thus causing the crank to twist, stretch and compress in other directions other than being cantilevered in the required direction or rotation.

Moreover, the pedal axis is parallel with the bottom bracket spindle but not necessarily normal to the neutral axis of the crank, so you may find that the bending measurement is not made in the same plane as the rotating pedal. Essentially, the best measurement cranks are with minimal geometric style and look like in-plane cantilever beams. Thus, for the purpose of the book, the back of the crank was machined to enable placement of gauges on a flat surface normal to the pedal axis (Fig. 5.16).

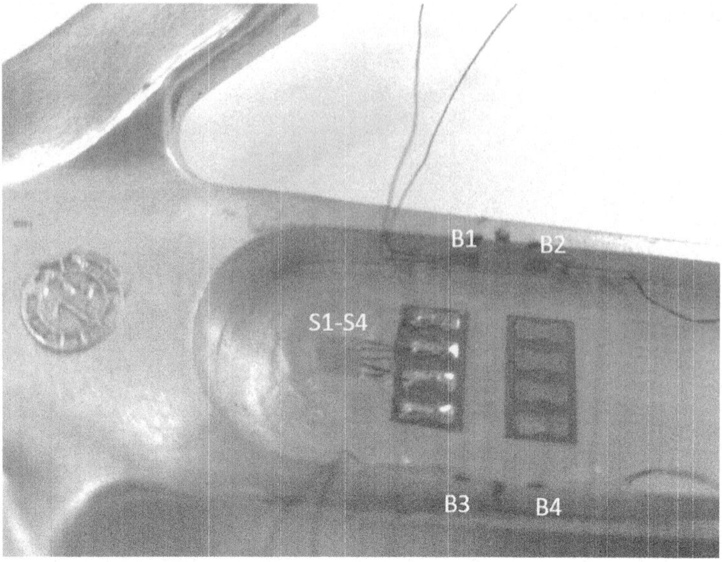

Fig. 5.16 Strain gauges to measure crank bending (*B*) by the cantilever effect with four single gauges and the maximum shear (*S*) stress along crank neutral axis by using an integrated element pattern

For investigatory purposes, the deflection of the crank was initially measured by placing gauges to measure the cantilever effect and the maximum shear strain along the neutral axis using a full Wheatstone bridge pattern as described on p. 92. The patterned gauge positioned to measure the shear strain suffered fewer crosstalks and are the results presented in this book. Because all the elements of the shear gauge are induced by lateral bending, by all being stretched or compressed, they cancel out the lateral bending crosstalk effect. As can be seen from Fig. 5.17, the elements of the gauge are aligned at 45° to the neutral axis and opposing elements pick up the positive or negative shear strain of crank bending in the required direction.

However, the shear strain gauges do pick up torque/twist crosstalk around the long axis of the crank; however, this effect was minimised by placing the gauge as close as possible to the torque axis by machining a channel into the crank. This helps because patterned gauge elements like this are designed to be placed on the outside surface of a shaft away from the torque axis to measure the shear strain of torque. As described on p. 102, to consider this crosstalk, the channel would have to be measured to be removed fully as a crosstalk.

The axial loading on the crank was measured again using a patterned gauge with two opposing elements that measured tension or compression and two elements that measured the Poisson effect (Fig. 5.18). The patterned element was placed at the pedal end of the crank so that lateral crank bending crosstalk strain was minimised.

The Poisson gauges are less sensitive (x Poisson ratio $\approx$ 0.3) than the gauges measuring tension or compression, and as a consequence, the axial gauge may not be

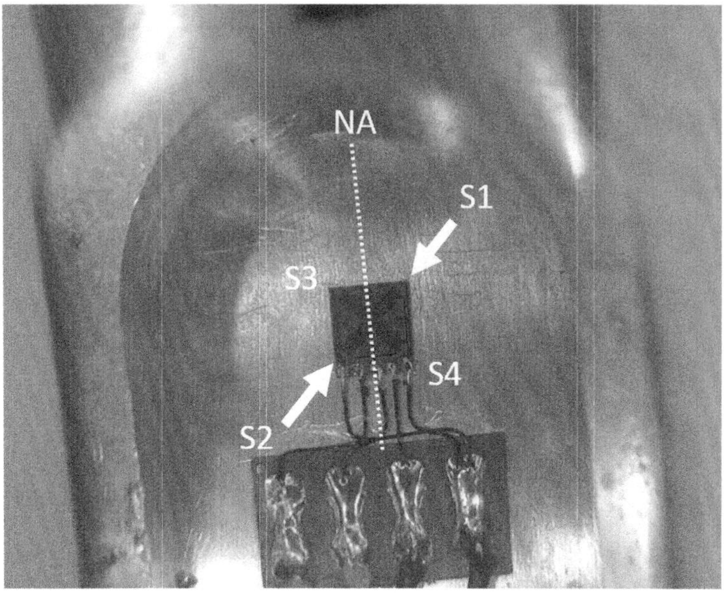

Fig. 5.17 Pattern gauge used to measure bending shear (*S*) along neutral axis (*NA*) with diagonally opposing gauges identified by white arrows

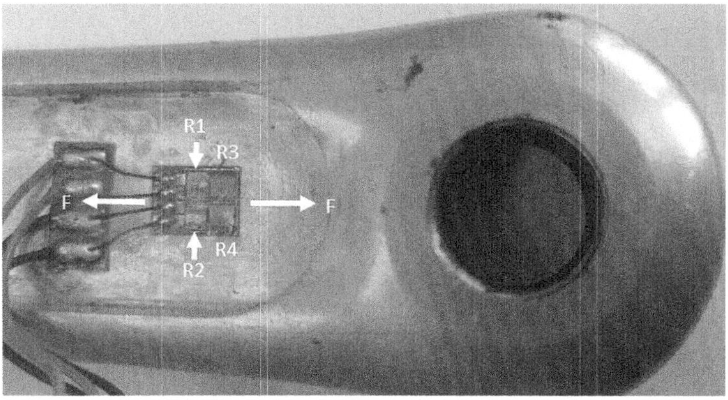

Fig. 5.18 The pattern gauge used to measure axial loading by measuring tension/compression and the passion effect. The diagonally opposing gauges identified by arrows are the passion gauges

as sensitive as the patterned gauge measuring shear strain at the bottom bracket end of the crank. This is seen by calibration and shown for the left crank; the radial strain gradient is −3871 and the bending shear is 5953 microvolt/force (Fig. 5.19).

Figures 5.20 and 5.21, respectively, show the power curve for the right and left leg and the effective and radial crank force for the left crank. The power curves are typical of that published in literature and illustrate good symmetry; the slight

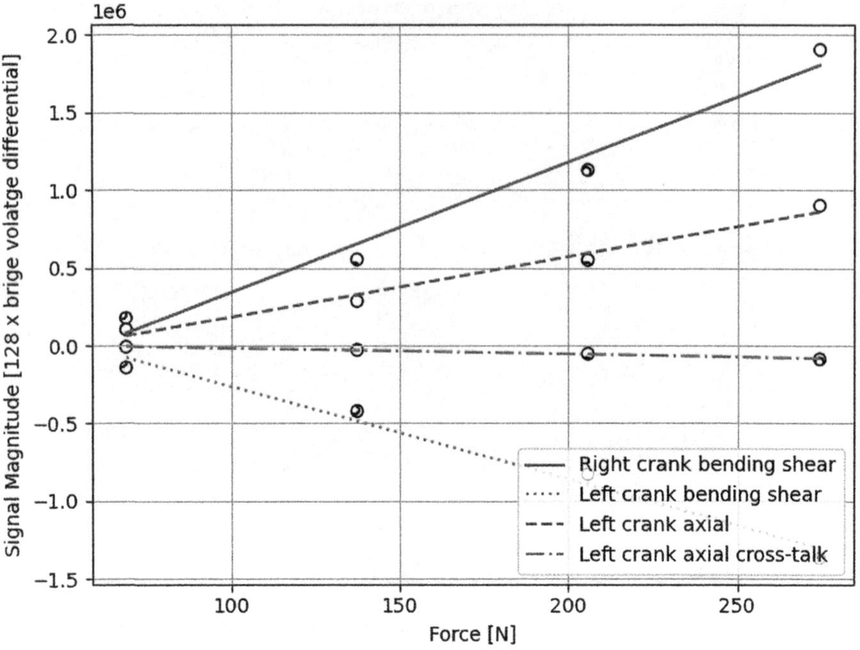

Fig. 5.19 Calibration curve using force plateaus against signal response like those shown on p. 102

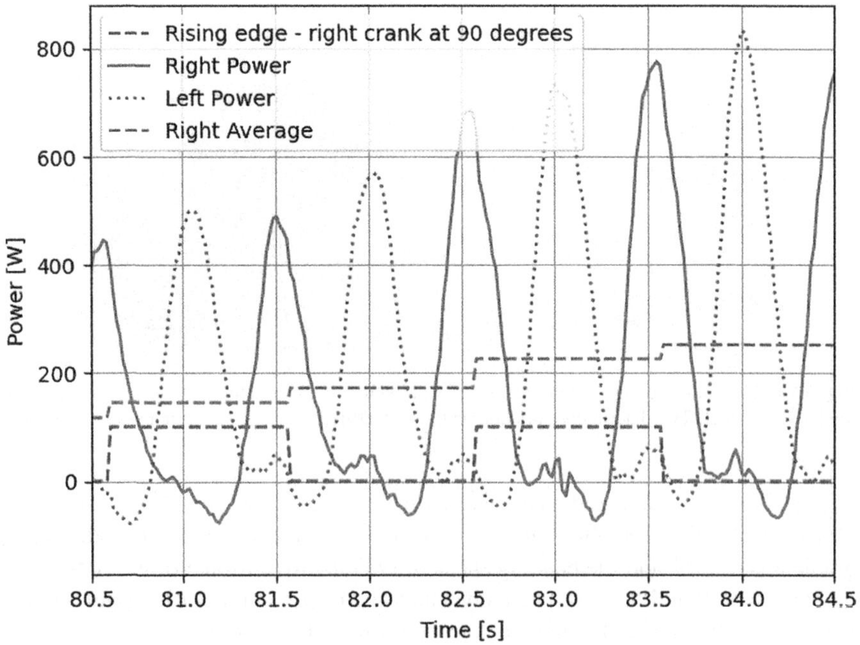

Fig. 5.20 Right and left leg effective power curve

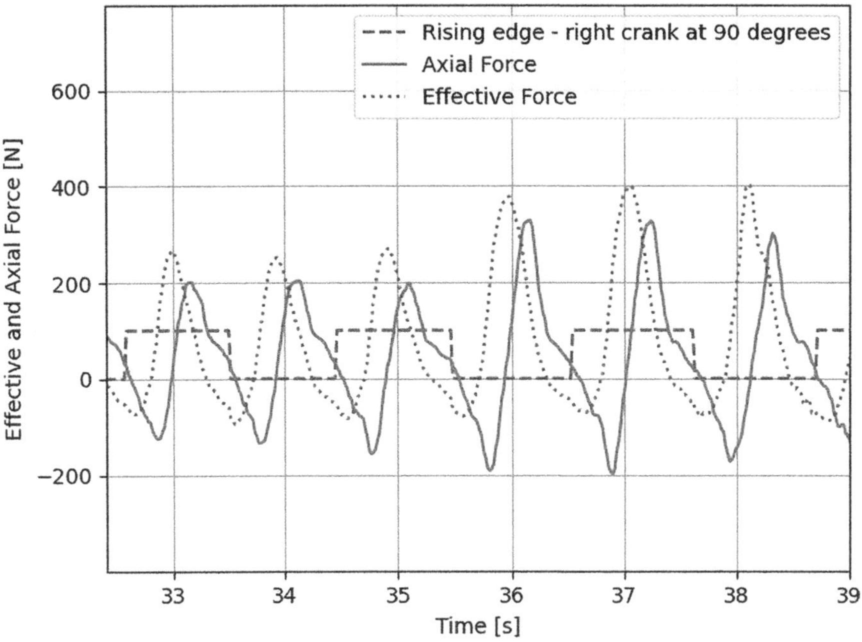

Fig. 5.21 Left crank effective and axial force

negative work done (as indicated by the negative power) is the work done by the working leg lifting the opposing leg on the upstroke and ideally should be at a minimum. The pulses of the power stoke display the power impulses of the muscles, and the average power developed is very much less than the maximum power developed.

The final plot shows that the radial crank force mimics the effective force and experiences tension and compression throughout the crank cycle.

5.4.2 Bottom Bracket Force Dynamometer

The main advantage in theory of the torque bottom bracket dynamometer is that the left and right limb power could potentially be recorded directly onto a single data stream. This could significantly reduce effort with respect to the electronics requirement, thus data logging synchronisation. However, the main limitation is positioning the gauge to measure either the combined left and right limb effort or even the individual limb effort.

Figure 5.22 illustrates why a single position cannot be used to measure the torque input from the left or right crank. Moreover, because of the position of the chainrings (at the 80 mm position in the graph), with respect to right- and left-hand pedal, the

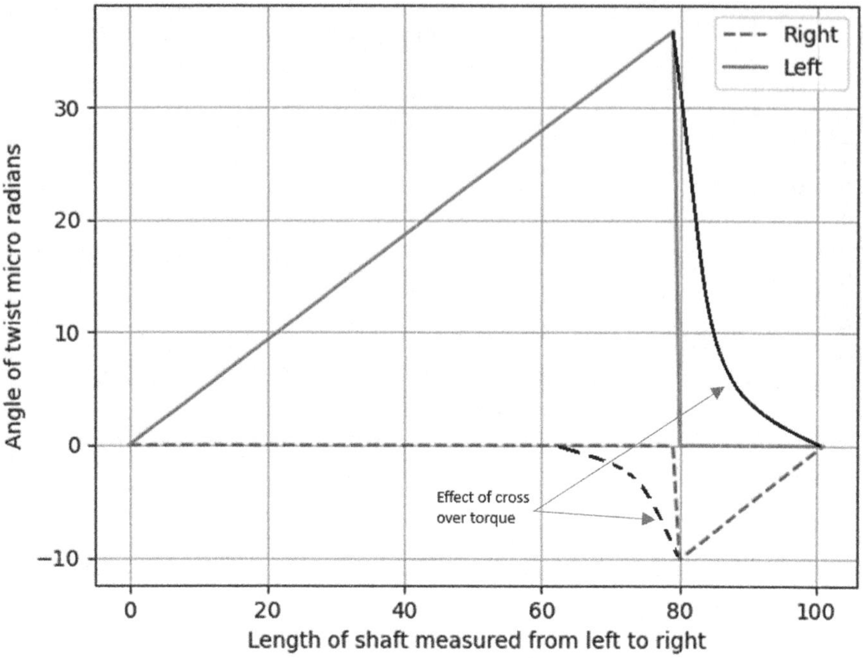

Fig. 5.22 An idealised schematic maximum variation of torque along the bottom bracket shaft 180° out of phase

shaft experiences less twist to the right of the chainrings, which causes sensitivity problems. For example, suppose took a cylindrical sponge of two lengths one longer than the other and drew a square on the surface say, at the fixed position of your left hand and twisted the sponge with your right hand. The square on the longer sponge will deform more than the square on the shorter sponge, even though the torque is the same. This also happens with the bottom bracket shaft when measuring the torque. Moreover, the torque applied by the left leg effects the right-hand measurement and vice versa; the torque drops off gradually after the chainring on its respective side. It was found that this type of measurement resulted in too much crosstalk, and only the left-hand limb input could be measured effectively. This sort of measurement may be possible by designing a new chainset component; although based on the efforts of retrofitting off-the-shelf components, it did not appear to be a worthwhile venture. Additionally, the bottom bracket shaft is also subject to bending and shear gauges positioned to measure torque and also the bending shear (90° to the direction of bending) of the rotating shaft which will be discussed later. Indeed, this can be minimised with good positioning although it is another challenge when considering direct torque measures.

However, the bending of the bottom bracket shaft can also be considered when measuring cyclist input. As shown by Fig. 5.23, the pedal forces result in a bending moment which is proportional to the force applied to the pedal. The difficultly of

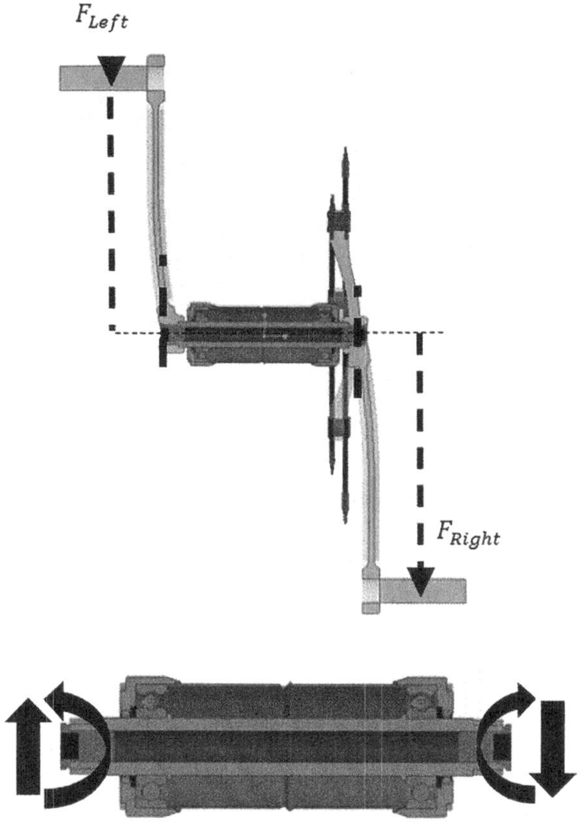

Fig. 5.23 The bending moment and force supported by the ends of the bottom bracket shaft

using the bending measurement is that it is not possible to correlate the bending signal with the separate input pedal force from the right leg and left leg. However, the reaction force applied by the chain is measured and is now discussed in greater detail.

As the bottom bracket shaft is rotating, its bending will occur in a single direction effected by the pull of the chain; by measuring the deformation magnitude with respect to a coordinate system, both the total magnitude and direction of bending will be known (Vuong et al. 2022). Figure 5.24 illustrates how two full-bridge Wheatstone bridges, R1-R4 and R5-R6, would be orientated to measure the direction of bending (Fig. 5.25).

The gauges were placed in line with the edges of a square tapered bottom bracket shaft and bonded so that they were located at the mid-point between the two bearings.

To calibrate the bottom bracket shaft, the easiest method is to fit the bottom bracket to the bicycle by orientating the cranks to essentially apply the effective force. As shown in Fig. 5.24, p. 113, circuit R1—R4 should only measure a change when the effective force is applied. However, the effect of the chain reaction force

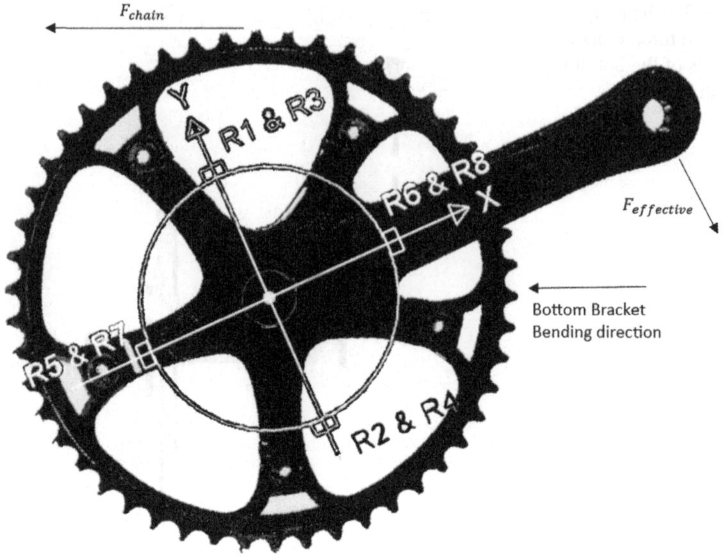

Fig. 5.24 The arrangement of two orthogonal Wheatstone circuits to measure applied crank force

(which is not crosstalk) could be seen in the orthogonal circuit (R5–R8), when setting the cranks to 90° (Fig. 5.26).

Thus, if the cranks are set to 0° (p. 52), and the weight acts vertically, the orthogonal channel should not be excited, as demonstrated in Fig. 5.27. This effect can be mitigated by building a jig so a chain is not used to hold the chainset in position. Fig. 5.28 is the signal/force calibration curve relationship of the measured weight plateaus.

However, the reader should be aware while it is convenient to load the cranks in this manner, it will not be as accurate as a bench test, and because of the chain reaction force, it is not possible to comment on crosstalks. This can be achieved by carrying out a four-point bending test where the bending moment is held constant between supports (Fig. 5.29).

One of the reasons for using the square taper bottom is the ease with which it is possible to use v-blocks to carry out four-point bending. The top v-block sits on two ridges (white arrows), and additional weights can be added on top (Fig. 5.30).

The calibration curve used in the example (p. 103) is taken from the calibration of the squared tapper shaft using four-point bending. As can be seen from a bench test, the "best-fit" curve does better match the recorded data points. However, while this calibration can be used to return the shaft bending moment, it is not possible to determine the crank effective force. However, the shaft is statically indeterminant as there are too many unknowns to determine the end boundary conditions as shown on p. 112, of the bottom bracket. More bending boundary conditions along the length (end conditions) of the bottom bracket shaft could only be measured by more strain

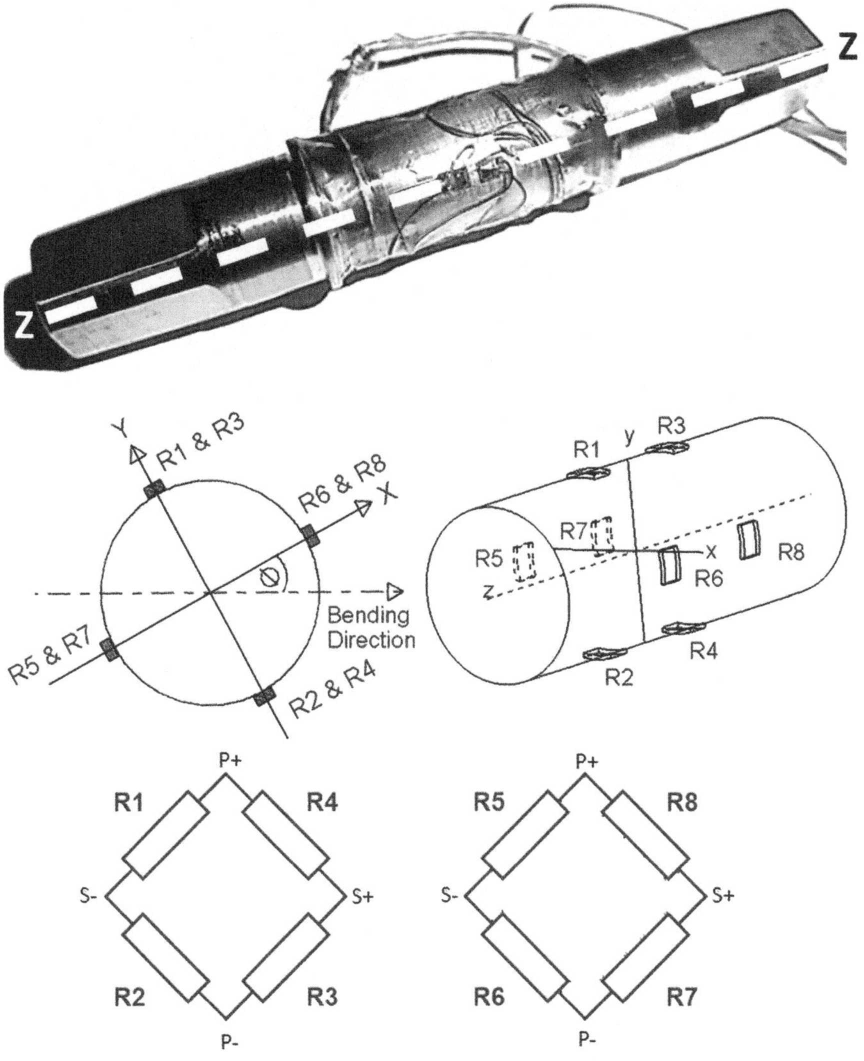

Fig. 5.25 Strain gauges positioned on the bottom bracket shaft with respect to crank orientation, *P* is power and *S* is signal

gauge circuits. Hence, the easiest method is to correlate a force applied on the cranks with the bottom bracket bending signal.

The reason for the bending calibration is that the orthogonality of the calibration matrix (C) can be checked (crosstalks) and the eigenvectors (V) returned. By spectral decomposition, the eigenvectors (V) from the four-point bending test can be decomposed with the two eigenvalue (λ) calibration constants, measured by applying forces to the cranks directly to return a full calibration matrix of crosstalks (Eqs. 5.20 and 5.21). The example shown Appendix 3, (code line 082), is the python

Fig. 5.26 Apply weights to the cranks in the 90° position for calibration

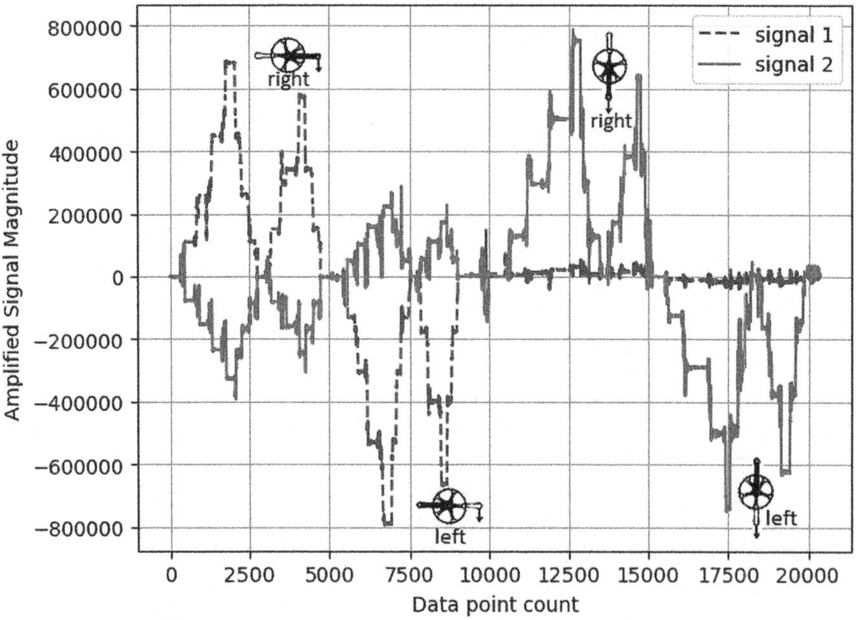

Fig. 5.27 Change in signal output for 7, 14, 21 and 28 kg mass added to calibration bars

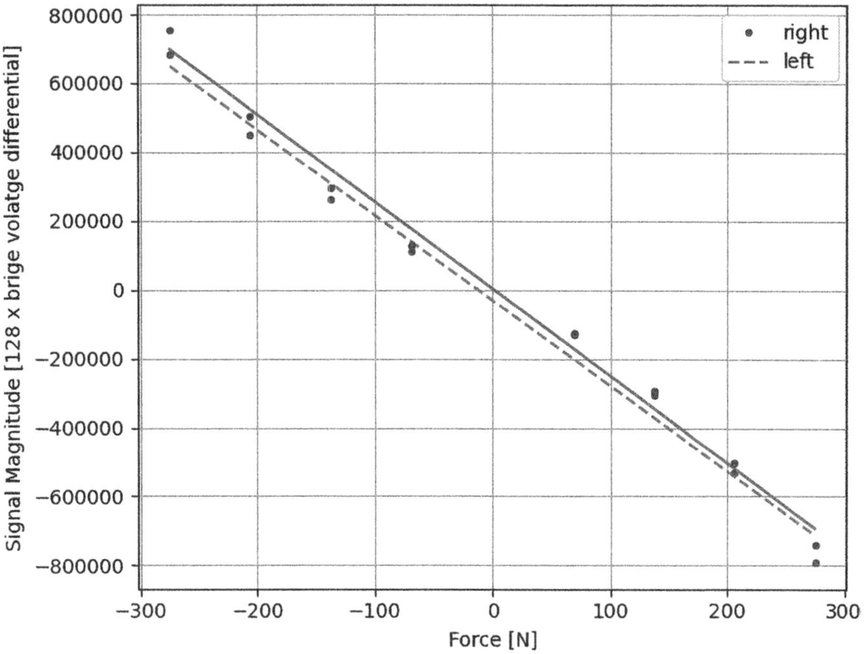

Fig. 5.28 Calibration curve using force plateaus against signal response, shown in Fig. 5.27

script showing the post processing of the bottom bracket data and this operation practically.

$$AX = V\lambda \tag{5.20}$$

$$\text{Calibration matrix } [C] = AX \tag{5.21}$$

As already indicated on Fig. 5.3, p. 118, the bending measured by the rotating orthogonal gauges pick up the bending magnitude (in the direction of the pull of the chain) with respect to their orientation. By treating the orthogonal signal magnitudes x and y, as the components of the force causing the shaft to bend in direction of the chain, it is possible to determine the orientation of the rotating cranks (Fig. 5.31). As discussed, Vuong et al. (2022) has documented the technique of using orthogonal gauges to determine the direction of bending propeller shafts. Figure 5.31 clearly demonstrates that as expected, the signal magnitudes denote the sinusoidal rotation of the cranks.

Hence, by knowing the orientation of the cranks with respect to the direction of bending, the magnitude of the force applied in the direction of the chain can be resolved. As expected, in the orthogonal direction to the chain, there is minimal force

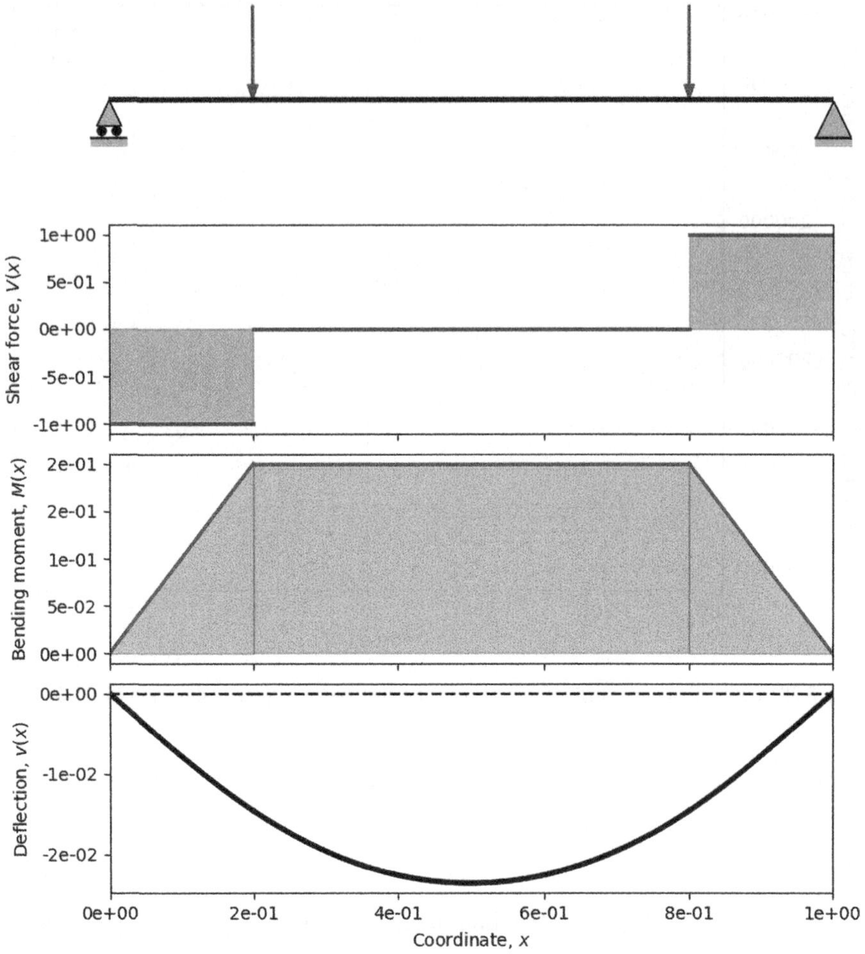

Fig. 5.29 Schematic of 100 mm long bottom bracket shaft undergoing four-point bending

applied, demonstrated by Fig. 5.32 for a short 2-min ride. Then as illustrated by Fig. 5.33, the power developed can be determined and represents a measurement closer to the average than the maximum and minimum power developed by the muscles. However, the fluctuations as developed by the crank revolutions can still be observed and demonstrates a novel method of power measurement while cycling. It should be noted that this technique is also more suited to single chainring use as the chain velocity is determined by taking the product of the chainring angular velocity and its diameter. If the double chainring is used, a sensor is required to return the chainring utilised at that moment in time.

Fig. 5.30 Four-point bending carried out on a squared tapered shaft. White arrows indicate ridges that top V-block sits on to inside of bearing placement

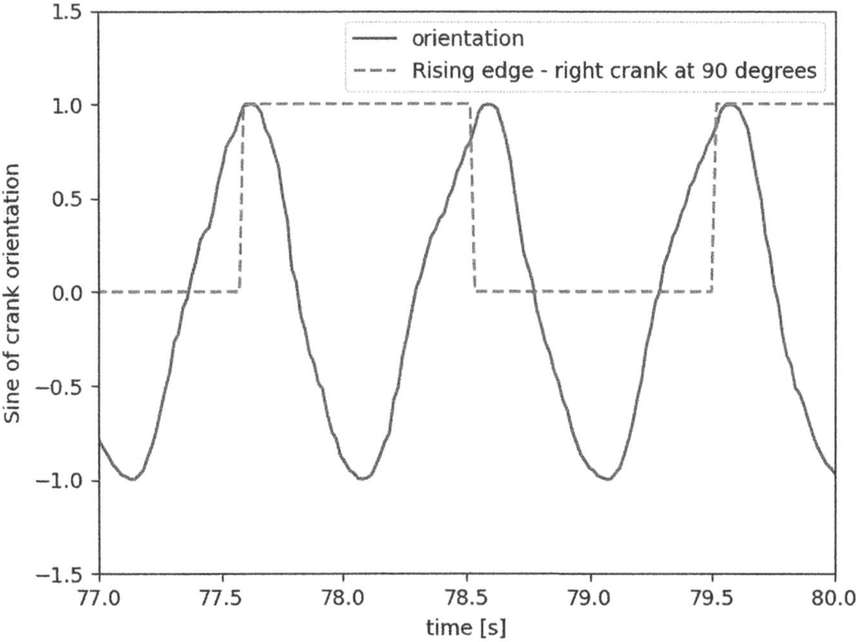

Fig. 5.31 Sine of the crank with respect to chain direction determined from eigenvector analysis of the two bending signals

5.4.3 Pedals Dynamometer

Pedal dynamometers have become more popular in recent years, as they are easy to fit to the bicycle, or cross over to the bicycle the rider may be using on that day. The

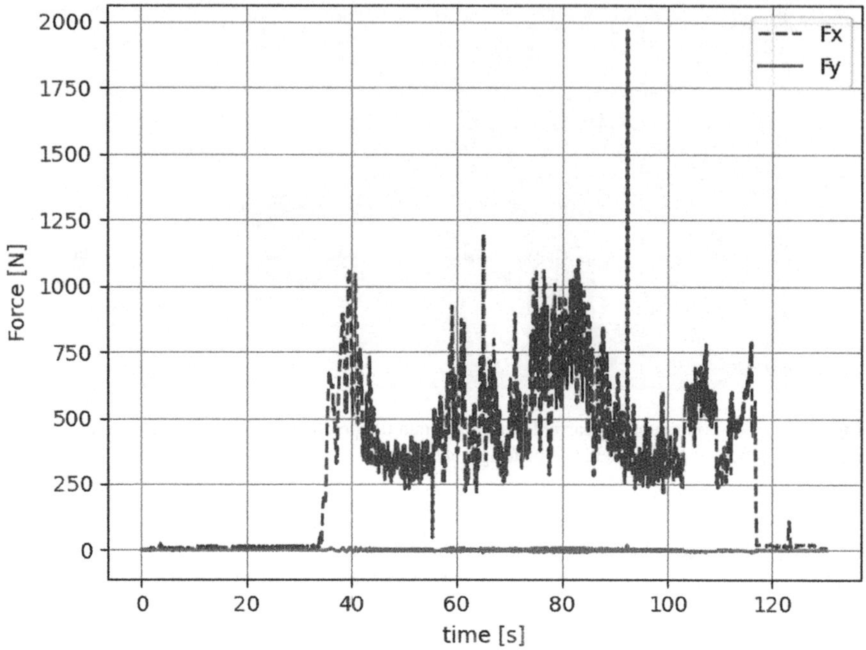

Fig. 5.32 The force applied to the chain (Fx) for a 2-min cycle

pedal shaft primarily experiences cantilever loading and suffers little crosstalk from unwanted deformations. Generally, it is reported in media that commercially available pedals have good accuracy compared to cranks; however, it cannot be concluded with a sweeping statement that pedal measurements are "better" than crank measurements, as it depends on the design—there are also poorly performing pedal systems.

However, the main difficulty with pedal designs is that there is little room for strain gauge placement and because threads start on spindles and in cranks in varying positions, the torqued orientation of the pedal shaft is unknown. Thus, the pedal force orientation has to be resolved: that is, orthogonal gauges fixed to the pedal spindle do not necessarily align with any meaningful axis. Another practical problem is that everything needs to be miniature; as the pedal rotates, it is not easy to hang a power source or an acquisition system off the crank. There are many solutions; however, if the reader is building a one-off pedal device for a specific crank, it is possible to mark the orientation of the pedal shaft (Fig. 5.34).

The back of the pedal shaft was machined, in this example as shown, and a miniature shear strain gauge was fitted, to align with the axis of the crank coordinate system (Fig. 5.35).

This allows a power source and data acquisition systems to be fitted to the crank and negates the requirement for the orientation of the forces to be resolved. The shear gauges used are shown more closely in Fig. 5.36; the orthogonal gauge to the axis in

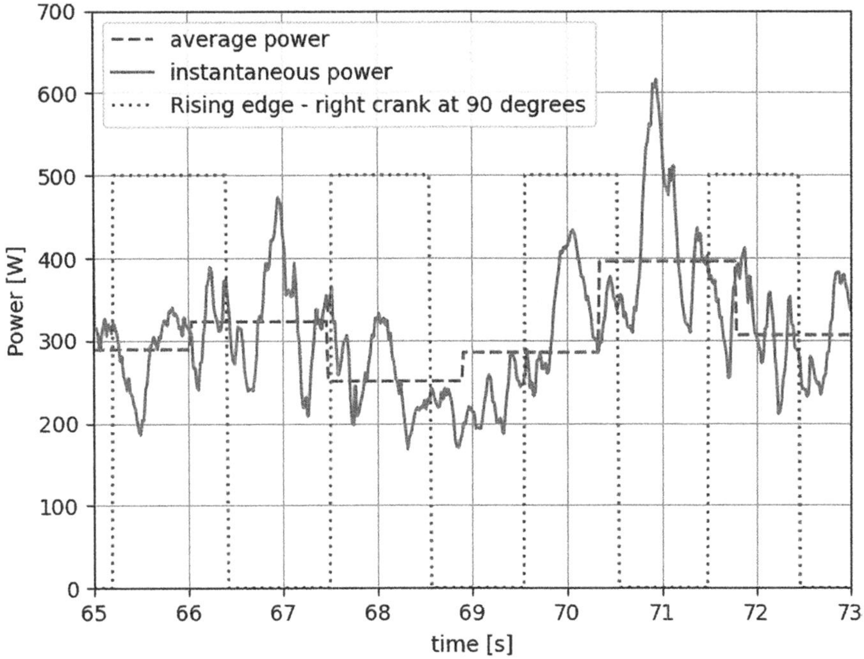

Fig. 5.33 Power developed by measuring the chain force

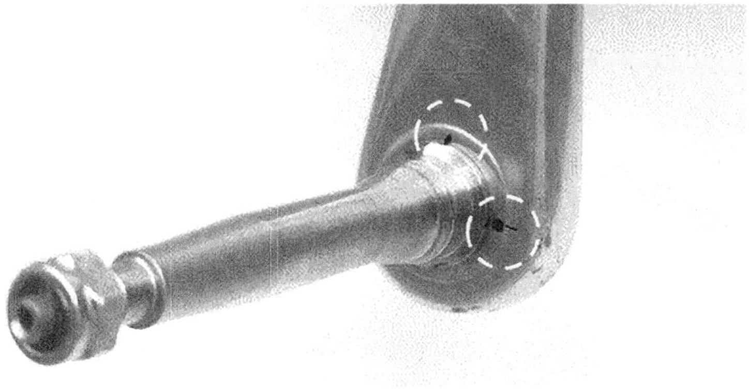

Fig. 5.34 Marking the orientation of the pedal shaft with respect to the crank

question measures the load in that direction. For example, the bending experienced in the y-direction shown in Fig. 5.35 is not measured by the shear gauge aligned in that plane (signal 2); instead, in the orthogonal position (signal 1), at the maximum shear, when a shaft is bending, it is orthogonal to the plane of bending. The patterned gauges are the same as those used for the cranks on p. 107, and because of their

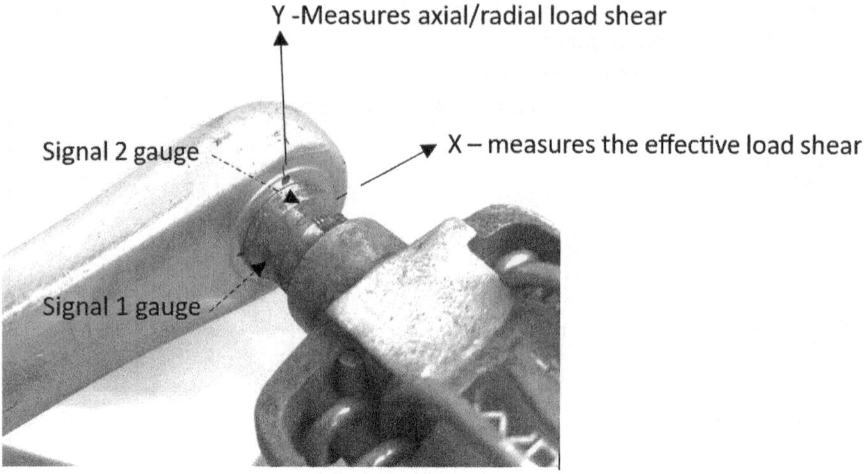

Fig. 5.35 Strain gauge placement on pedal shaft

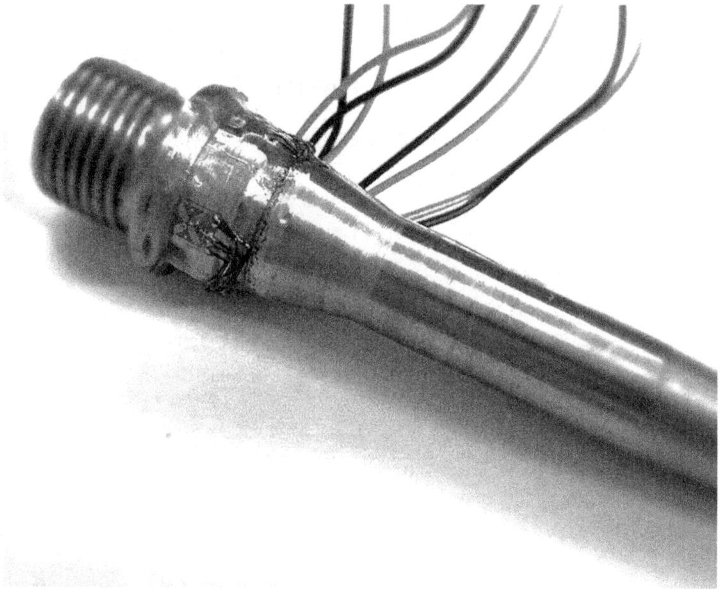

Fig. 5.36 Shear pattern gauge to measure pedal force

orientation, if the pattern gauge experienced bending, all the elements would be excited and no voltage differential would be recorded.

As demonstrated by Fig. 5.37, the shear signal induced by hanging weights off the pedal shaft in the Y-direction results in the gauge in the orthogonal position measuring shear (Signal 1)—there is minimal crosstalk experienced by the gauge in the plane of bending.

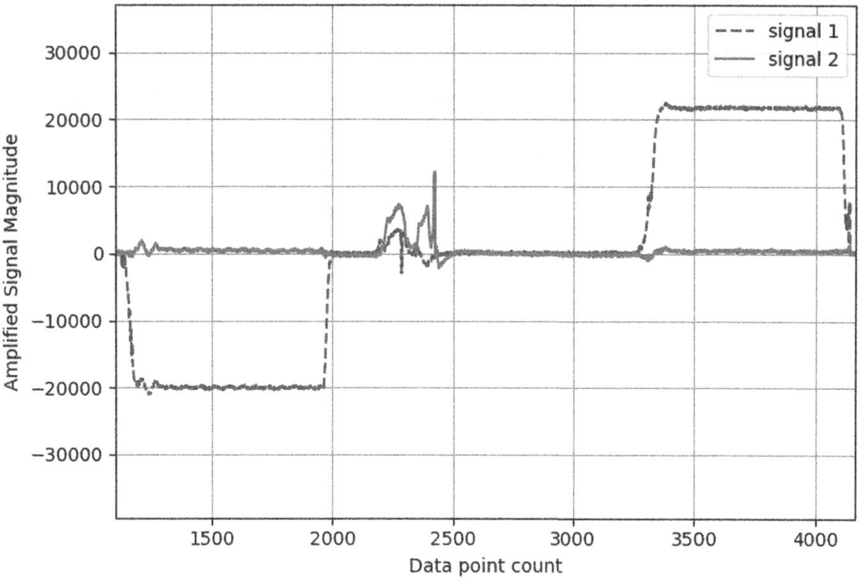

Fig. 5.37 Pedal shear force calibration achieved by hanging weights of the pedal shaft in the Y-direction

Another advantage of measuring the shear force is that the position of the point loads or uniformly distributed loads, respectively, applied by either radial or roller bearings do not need to be considered (Fig. 5.38). This is because of the shear force measured at the fixed end of the cantilever is equal to the sum of all the cantilever loading (Fig. 5.38). On the other hand, if bending is measured, the boundary conditions and therefore the positions of the loads and their magnitudes should be known. Hence, by hanging a weight along the pedal shaft for shear calibration, the position of the weight is not critical although it should be repeatedly placed, as this is good practice.

The calibration, as discussed in previous sections of this chapter, can be achieved by treating the two signals as individual components or as a matrix of cross talks—the latter was adopted to present the measured result in Fig. 5.39.

The effective force plot is similar to that made on the crank-based system shown on p. 110—although it should be noted that the rising edge of the square wave denotes a position when the right crank is at 270° (p. 52) compared to the example on p. 110 where the left crank is at the 270° position.

The measurement of the effective force is clearly of the same shape (p. 110) and demonstrates that during the upstroke, the force applied to the pedal opposes the direction of rotation. Additionally, the output clearly validates the two methods of measurement, although in the author's opinion, the measurement of the pedal shaft force compared to the crank force resulted in less crosstalk between loading directions.

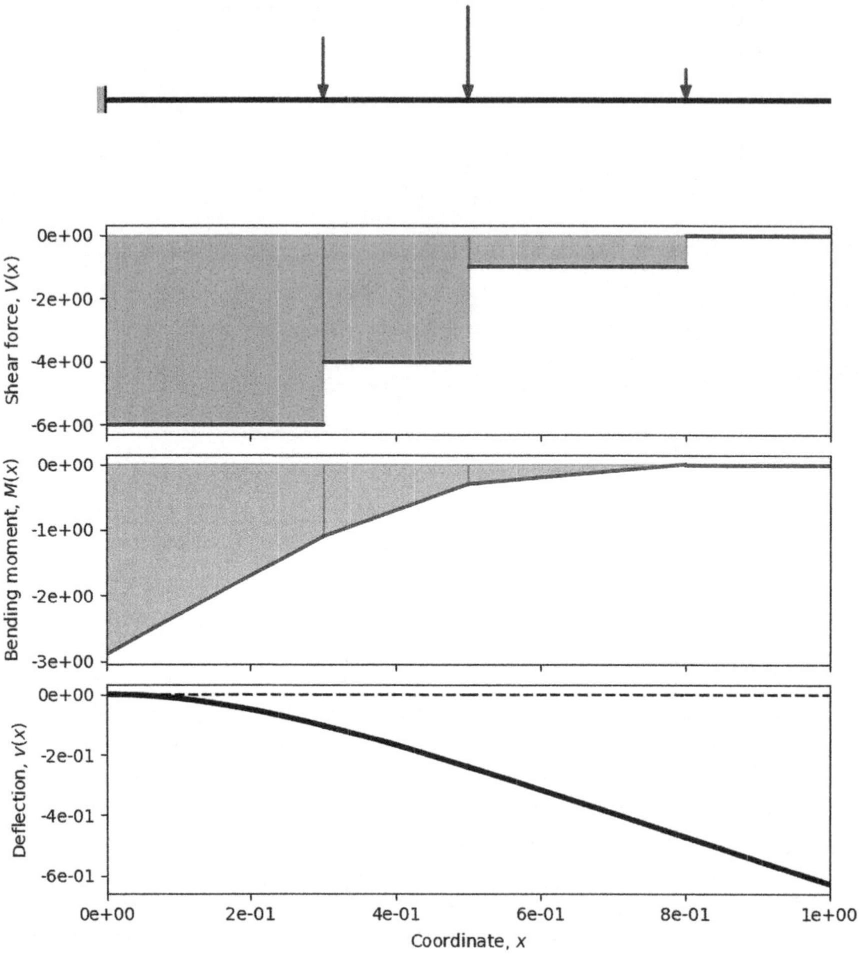

Fig. 5.38 Bending of a cantilever beam; the shear at the fixed end is equal to the sum of the forces acting along the cantilever

On a final note, the reader my question why the bottom bracket dynamometer was not built using a similar technique (shown for the pedals) by measuring the shear of bending. As already mentioned on p. 110, the bottom bracket shaft also experiences torque, and to position the gauges to measure only the shear bending stress and not torque or vice versa is challenging, and to that end, it was easier to measure the bending directly.

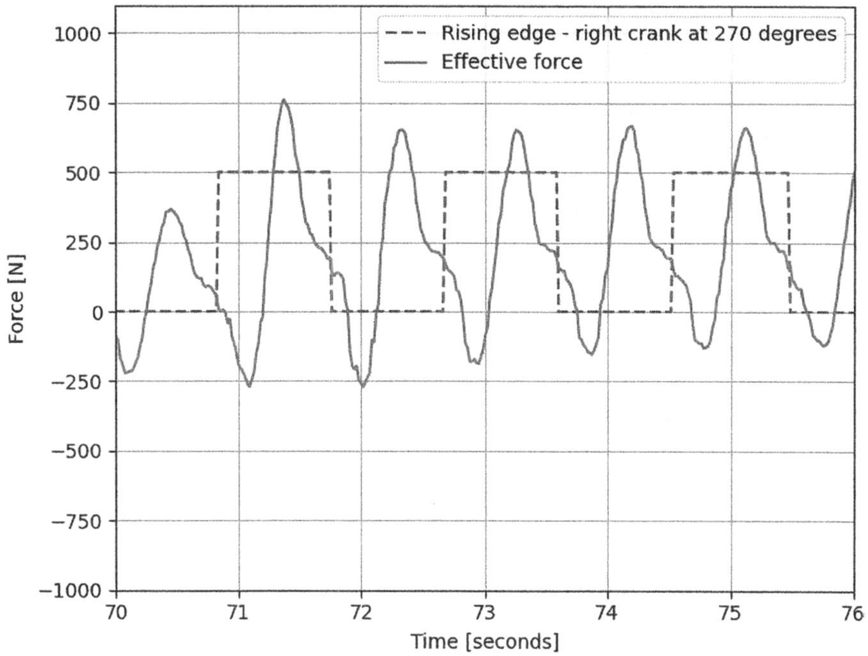

Fig. 5.39 The effective force measured by the pedal shaft with respect to the crank

5.5 Conclusion

This chapter has highlighted some considerations when buying a power measurement dynamometer from a technical point of view. Most commercial dynamometers are for power cadence setting, and as a training tool, this method of cadence training is commonly utilised. However, to analyse your biomechanics by evaluating the symmetry and style of pedalling in detail, a time series trace is required, especially when considering the resting leg in the upstroke or understanding muscle impulse development. Indeed, by objectively measuring human motion, it provides the engineer, inventor and enthusiast with an objective insight into the interaction of the rider with their machine. The strain gauging techniques shown in this book can be adopted for fully instrumenting the bicycle to measure for example, the handle bar and seat loading.

There are commercially available systems for power measurement, though they are expensive, and you still may only have limited insight into how they are functioning. However, it may not necessarily be cost-effective to build your own dynamometer although it has other benefits such as being able to objectively quantify the accuracy and precision of your device. The bottom bracket measurements presented in this book illustrate a novel measurement method that may be useful for returning the average power directly when cycling—which may be more useful than observing the sinusoidal input by the muscles.

While this book has primarily concentrated on the road bicycle set-up, it is hoped that this publication may act as spring board for the reader interested or kindle the interest of measuring and understanding the kinematics and kinetics of human power vehicles. The bicycle was developed iteratively by understanding human factors, physiology and biomechanics and is the collective design and inspiration of many inventors—and there is still opportunity to become part of its story.

Bibliography

BSSM. British Society for Strain Measurement. From https://www.bssm.org/

McAinsh R (2014) Design and engineering of an accurate bicycle power meter. ICSST 2014, Singapore. https://www.researchgate.net/publication/303517617_Design_and_Engineering_of_an_Accurate_Bicycle_Power_Meter/references

Rowe T, Hull ML, Wang EL (1998) A pedal dynamometer for off-road bicycling. J Biomech Eng 120(1):160–164

Vuong QD, Lee W-J, Choi H, Seo K, Kim Y, Jeong JH, Song M-H, Lee J-U (2022) Establishing the true dynamic bending moment of propeller shaft using a single bridge of strain gauge. Appl Sci 12(18):9235

Appendices

A. Crimin, A. McGarry, *Bicycle Biomechanics*,
https://doi.org/10.1007/978-3-031-85712-6

Appendix 1

Five-Bar Mechanism Model

```
import matplotlib.pyplot as plt
import numpy as np
from mechanism import *
from scipy import interpolate

# Declare the joints that make up the system.
O, PEDAL, ANKLE, KNEE, HIP = get_joints('O PEDAL ANKLE KNEE HIP')

PEDAL.follow = True
KNEE.follow = True
ANKLE.follow = True

# https://msis.jsc.nasa.gov/sections/section03.htm
# Body Size of the 40-Year-Old American Male for Year 2000 in One Gravity
Conditions
# 50th percentile
# The Anatomy and Biology of the Human Skeleton (Steel and Bramblett 1988).
# For entire humerus:  (3.08 x length of humerus) + 70.45 = stature +/- 4.05 cm


length_of_humerus = 40.4
Tibiale_height = 41.1
saddle_height = 80
ball_of_foot = 19.3
crank_length=17.2


# Declare the vectors and keep in mind that angles are in radians and start from
the positive x-axis.
crank_pedal = Vector((O, PEDAL), r=crank_length)
ankle_pedal = Vector((PEDAL, ANKLE), r=ball_of_foot)
ankle_knee = Vector((ANKLE, KNEE), r=Tibiale_height)
knee_pedal = Vector((PEDAL, KNEE), show=False)
knee_hip = Vector((HIP, KNEE), r=length_of_humerus)
hip_crank = Vector((O, HIP), r=saddle_height, theta=(np.pi/180)*108,
style='ground')

# Define the known input to the system.
# For a 60rpm
rpm = 60
T = 60/rpm  # period s
n = 200 # 200 data points simulated for revolution
t = np.linspace(0, T, n)

angular_velocity = -2*np.pi*1/T  # This is RPM in rad/s
theta_a = angular_velocity*t + np.pi/2 # Integrate to find the angle


#ankle angle measure by Kinovea software
ankle_angle = np.array([127.2944946, 124, 122, 120.03884120000001,
117.14366909999998, 117.1700668, 121.69084170000001, 129.2927246, 137.34970090000002,
143.2788544, 145.69099430000003, 145.3253784, 143.6055145, 141.48645019999998,
140.2263031, 140.8285675, 142.2711182, 142.2412415, 139.6714325, 135.37277219999999,
131.2464905, 128.6952972, 127.2944946])*np.pi/180

#interpolating angle angle with cubic spline to create 200 data points
f =
interpolate.interp1d(np.linspace(0,len(ankle_angle),len(ankle_angle)),ankle_angle,
kind='cubic')
#theta_b is used to constrain the ankle angle
theta_b = f(np.linspace(0,len(ankle_angle),n))

#the angle angle as a sine function (uncomment to use)
#theta_b = 150*np.pi/180+np.deg2rad(8)*np.sin(angular_velocity*t)

#ankle angle set to 90 degrees
#theta_b  = np.repeat(0.5*np.pi,len(t))

```

```
knee_to_pedal = (ankle_pedal.pos.r**2 + ankle_knee.pos.r**2 - 2* ankle_pedal.pos.r * ankle_knee.pos.r*np.cos(theta_b))**0.5
knee_to_pedal_dot = np.gradient(knee_to_pedal, t, edge_order=1)
knee_to_pedal_2dot = np.gradient(knee_to_pedal, t, edge_order=2)

omega = np.full((t.size,), angular_velocity)  #An array of the same angular velocity
alpha = np.zeros(t.size)


# Guess the unknowns (values at t=0)
pos_guess = np.deg2rad([135, 60, -30, 90])  # at least get it in the right quadrant
vel_guess = np.array([1000, 1000, 1000, 1000])
acc_guess = np.array([1000, 1000, 1000, 1000])

# Define the loop equation(s)
def loop(x, i):
    temp = np.zeros((2, 2))
    temp[0] = knee_hip(x[2]) - ankle_knee(x[1]) - ankle_pedal(x[0]) - crank_pedal(i[0]) + hip_crank()
    temp[1] = ankle_pedal(x[0]) + ankle_knee(x[1]) - knee_pedal(i[1], x[3])

    return temp.flatten()


# Create the mechanism object
mechanism = Mechanism(vectors=(crank_pedal, ankle_pedal, ankle_knee, hip_crank, knee_hip, knee_pedal), origin=O,loops=loop, pos=np.stack((theta_a, knee_to_pedal), axis=1), vel=np.stack((omega,knee_to_pedal_dot), axis=1), acc=np.stack((alpha, knee_to_pedal_2dot), axis=1), guess=(pos_guess,vel_guess, acc_guess))


# Call mechanism.iterate() then get and show the animation
angles = np.linspace(0, 360, t.size)  # Use this array for stamp and x-axis graphs
mechanism.iterate()
ani, fig_, ax_ = mechanism.get_animation(velocity=True, show_joints=True, cushion=5, interval=0,
                                          stamp=angles, stamp_loc=(0.65, 0.9))


# Plot the angles, angular velocity, and angular acceleration of vector d
fig, ax = plt.subplots(nrows=3, ncols=1)

th2 = -knee_hip.pos.thetas
th1 = np.pi-(ankle_knee.pos.thetas-knee_hip.pos.thetas)
#th3 = ankle_pedal.pos.thetas


ax[0].plot(angles, np.rad2deg(th2), color='maroon')
ax[1].plot(angles, np.rad2deg(th1), color='maroon')
ax[2].plot(angles, np.rad2deg(theta_b), color='maroon')


ax[0].set_ylabel("Hip Angle [$\\alpha$] \n<-Flexion -- Extension->")
ax[1].set_ylabel("Knee Angle [$\\beta$] \n<-Extension -- Flexion->")
ax[2].set_ylabel("Ankle Angle [$\\gamma$] \n<-Dorsi -- Plantar-flexion->")
ax[2].set_xlabel(r'Crank Angle [$\theta$]')


for ankle_pedal in (ax[0], ax[1], ax[2]):
    ankle_pedal.minorticks_on()
    ankle_pedal.grid(which='both')
fig.set_size_inches(7, 7)


plt.savefig("thigh and shank angle 5 bar")
```

```
#recorded force data
#vertical pedal force (local)
#vertical force  - points interpolated to fit 200 data points
xv = np.array([  0. ,  29.3,  45. ,  76.6,  90. , 102.4, 124.7, 140.9, 162.2,
       189.8, 225. , 255.5, 270. , 315. , 360. ])
yv = np.array([   0.  ,  170.  ,  317.22,  510.  ,  601.12,  649.57,  510.  ,
        340.  ,  170.  ,    0.  , -141.95, -170.  , -173.74,  -91.97,
          0.  ])


f = interpolate.interp1d(xv, yv, kind='cubic')
F_mag_V = f(angles)
#horizontal force - points interpolated to fit 200 data points
xh = np.array([  0. ,  22.9,  45. ,  52.1,  78.1,  95.3, 111.2, 135. , 180. ,
       225. , 270. , 292.7, 315. , 360. ])
yh= np.array([   0.  , -170.  , -199.24, -170.  ,    0.  , 170.  , 340.  ,
        443.02,  403.92,  214.03,   68.85,    0.  ,  -48.45,  -70.72])
f = interpolate.interp1d(xh, yh, kind='cubic')
F_mag_H  = f(angles)



#data export
#pedal, ankle knee and hip
x_p = PEDAL.x_positions
y_p = PEDAL.y_positions
p = [i for i in zip(x_p,y_p)]

x_a = ANKLE.x_positions
y_a = ANKLE.y_positions
a = [i for i in zip(x_a,y_a)]

x_k = KNEE.x_positions
y_k = KNEE.y_positions
k = [i for i in zip(x_k,y_k)]

x_h = HIP.x_positions
y_h = HIP.y_positions
h = [i for i in zip(x_h,y_h)]

def matrix_define(o1,o2,x1,y1):
    #enter value for long mechanical axis first
    #o1 x origin of system
    #o2 y origin of system
    #x1 distal postion of a joint
    #y1 distal postion of the same joint

    #this can be at any angle with respect to long axis
    #o3 x origin of system
    #o4 y origin of system
    #x2 proximal postion of a joint
    #y2 proximal postion of the same joint

    x_1 = x1-o1
    x_2 = y1-o2

    #local axes along foot X_ is temporary
    X = [[(ii/np.linalg.norm(ii))] for i, ii in enumerate(zip(x_1,x_2))]


    X = [ np.concatenate((i[0],[0])) for i in X]# adding z dimension

    Y = [[-x[1],x[0],0] for x in X]# y axis pointing in direction of crank motion
    Z = np.cross(Y,X)

    matrix = [ np.matrix([[x[0],y[0],0],[x[1],y[1],0],[0, 0, z[2]]]) for x, y, z
in zip(X,Y,Z)]

```

```
    #so that for can be transposed to global coordinates
    matrix_inv = [np.transpose(i) for i in matrix]

    return matrix, matrix_inv

matrix_crank, matrix_inv_crank= matrix_define(0,0,x_p,y_p)
#Local measured pedal force
PF = [[x,y,0] for x,y in zip(F_mag_H,F_mag_V)]

#pedal force in global coordinates
PF_global = [np.matmul(matrix_inv_crank[i],ii) for i, ii in enumerate(PF)]

#ankle moment, angular velocity & power
m_a = [np.cross(ii,[p[i][0]-a[i][0],p[i][1]-a[i][1]]) for i, ii in enumerate(PF_global)]
Wt_a = np.gradient(theta_b,t)
p_a = [ii*i*-0.001 for i, ii in zip(m_a,Wt_a)]
p_a = [i[0][2] for i in p_a]

#knee [0][0,2]
m_k = [np.cross(ii,[p[i][0]-k[i][0],p[i][1]-k[i][1]]) for i, ii in enumerate(PF_global)]
Wt_k = np.gradient(th1,t)
p_k = [ii*i*0.001 for i, ii in zip(m_k,Wt_k)]
p_k = [i[0][2] for i in p_k]

#hip [0][0,2]
m_h = [np.cross(ii,[p[i][0]-h[i][0],p[i][1]-h[i][1]]) for i, ii in enumerate(PF_global)]
Wt_h = np.gradient(-th2,t)
p_h = [ii*i*0.001 for i, ii in zip(m_h,Wt_h)]
p_h = [i[0][2] for i in p_h]


#pedal_power
pedal_power = [f*0.170*-angular_velocity for f in F_mag_V]
# Plot the angles, angular velocity, and angular acceleration of vector d
fig, ax = plt.subplots(nrows=4, ncols=1)

ax[0].plot(angles, pedal_power, color='maroon')
ax[1].plot(angles, p_h, color='maroon')
ax[2].plot(angles, p_k, color='maroon')
ax[3].plot(angles, p_a, color='maroon')

#ax[0].set_title("Joint Power")

ax[0].set_ylabel(r'Pedal power [W]')
ax[1].set_ylabel(r'Hip power [W]')
ax[2].set_ylabel(r'Knee power [W]')
ax[3].set_ylabel(r'Ankle power [W]')
ax[3].set_xlabel(r'Crank Angle $\theta$')

for ankle_pedal in (ax[0], ax[1], ax[2], ax[3]):
    ankle_pedal.minorticks_on()
    ankle_pedal.grid(which='both')

fig.set_size_inches(8, 7)
plt.savefig("joint power")

#pedal vector plot
fig, ax = plt.subplots()
fig.set_size_inches(8, 7)
cont = 10
ax.quiver(x_p[::cont ],y_p[::cont ],-F_mag_H[::cont ],-F_mag_V[::cont ], scale=400, units="xy")
ax.plot(0,0,"o")
ax.set_xlabel("x direction [mm]")
ax.set_ylabel("y direction [mm]")
```

```
plt.savefig("vector plot")

#resultant force vesus tangental force
fig, ax = plt.subplots()
fig.set_size_inches(8, 7)
Resultant = [np.linalg.norm(i) for i in PF]
ax.plot(angles,Resultant)
ax.plot(angles,F_mag_V,"k:")
ax.plot(angles,F_mag_H,"k--")
ax.set_xlabel(r'Crank Angle [$\theta$]')
ax.set_ylabel('Pedal Force [N]')
ax.legend(["Resultant Force","Tangential force","Axial Force"])
ax.grid()
plt.savefig("crank tangential and axial force")

#polar power plot
ax = plt.subplot(111, projection='polar')
fig.set_size_inches(8, 7)
ax.plot(angles* np.pi / 180, p_h, 'k--', linewidth=3)
ax.plot(angles* np.pi / 180, p_k, 'k.', linewidth=3)
ax.plot(angles* np.pi / 180, p_a, 'k:', linewidth=3)
ax.plot(angles* np.pi / 180, pedal_power, 'r', linewidth=3)
ax.legend(["Hip Power [watts]","Knee power [watts]","Ankle power [watts]","Pedal
power [watts]"],loc=(0.8,1))
ax.set_theta_zero_location("N")
ax.set_theta_direction(-1)
ax.xaxis.set_minor_formatter('watts')
#ax.set_yticklabels([])
plt.savefig("polar plot of muscle power")
plt.show()
```

Appendix 2

Arduino Data Capture Code

```
#include "HX711.h"
#include <SPI.h>
#include <SD.h>

const int chipSelect = SDCARD_SS_PIN;
HX711 scale_ch1;
HX711 scale_ch2;

uint8_t dataPin1 = 6;
uint8_t clockPin1 = 7;

uint8_t dataPin2 = 8;
uint8_t clockPin2 = 9;

//uint32_t start, stop;
volatile float f1;
volatile float f2;

unsigned long myTime;
unsigned long lastInterrupt;
unsigned long period;
const byte interruptPin = 0;
int Flag = 0;//to stop frequency in interrupt being updated more than once

void setup() {
  // open the serial port at 9600 bps:
  Serial.begin(115200);
  pinMode(interruptPin, INPUT_PULLUP);
  attachInterrupt(digitalPinToInterrupt(interruptPin), blink, LOW);

  scale_ch1.begin(dataPin1, clockPin1);
  scale_ch2.begin(dataPin2, clockPin2);

  scale_ch1.set_scale(1);       // TODO you need to calibrate this yourself.
  scale_ch2.set_scale(1);       // TODO you need to calibrate this yourself.
  // reset the scale_ch1 to zero = 0
  scale_ch1.tare(2000);
  scale_ch2.tare(2000);

  //// setting up card
   while (!Serial) {
    ; // wait for serial port to connect. Needed for native USB port only
  }

```

```
  Serial.print("Initializing SD card...");

  // see if the card is present and can be initialized:
  if (!SD.begin(chipSelect)) {
    Serial.println("Card failed, or not present");
    // don't do anything more:
    while (1);
  }
  Serial.println("card initialized.");

  // set file headers
  // open the file. note that only one file can be open at a time,
  // so you have to close this one before opening another.
  File dataFile = SD.open("datalog.txt", FILE_WRITE);

  if (dataFile) {
    //the apps will measure time if Flag = 1 denotes revolution
    dataFile.print("crank period 0 or 1");
    dataFile.print(",");
    dataFile.print("time interval [ms]");
    dataFile.print(",");
    dataFile.print("Normalised force 1");
    dataFile.print(",");
    dataFile.println("Normalised force 2");
    dataFile.close();
    // print to the serial port too:
    //Serial.println(dataString);
  }
  // if the file isn't open, pop up an error:
  else {
    Serial.println("error opening datalog.txt");
  }


    // initialize digital pin LED_BUILTIN as an output.
  pinMode(LED_BUILTIN, OUTPUT);
  digitalWrite(LED_BUILTIN, HIGH);  // turn the LED on (HIGH is the voltage level)
}

void loop() {


  f1 = scale_ch1.get_units(1);
  f2 = scale_ch2.get_units(1);

```

```
  Serial.print(Flag); //the apps will measure time if Flag = 1 denotes revolution
  Serial.print(",");
  Serial.print(period);
  Serial.print(",");
  Serial.print(f1);
  Serial.print(",");
  Serial.println(f2);


  // open the file. note that only one file can be open at a time,
  // so you have to close this one before opening another.
  File dataFile = SD.open("datalog.txt", FILE_WRITE);


  // if the file is available, write to it:
  if (dataFile) {
    //dataFile.println(dataString);
    dataFile.print(Flag); //the apps will measure time if Flag = 1 denotes revolution
    dataFile.print(",");
    period = millis()-myTime;
    myTime = millis();
    dataFile.print(period);
    dataFile.print(",");
    dataFile.print(f1);
    dataFile.print(",");
    dataFile.println(f2);
    dataFile.close();
    // print to the serial port too:
    //Serial.println(dataString);
  }
  // if the file isn't open, pop up an error:
  else {
    Serial.println("error opening datalog.txt");
  }
}

void blink() {
  if(millis() - lastInterrupt > 5) // we set a 5ms no-interrupts window
    {
    if (Flag==1) {
    Flag=0;
    } else {
    Flag=1;
    }
    lastInterrupt = millis();
    }
```

Appendix 3

Python Post-processing Bending Bottom Bracket Dynamometer Script

```
import numpy as np
import matplotlib.pyplot as plt
from scipy import signal
import sympy as sym
from sympy import solve
from sympy import symbols
import math

def RMS(y):
    y=np.array(np.mean(y))**2
    return np.sqrt(y)


def Rising_edge(S1,S2,revolutions):

    # x 100 so the sqaure wave varies between 0-100 so it can be seen on the graph
    pedal_position = [i for i in revolutions]

    #using the rising and falling edge of the sqaure wave formed by the crank reed
    #switch passing an magnet. The average velocity for each pedal revolution is
    #returned
    t0 = time[0]
    t1 = 0
    rpm = np.zeros(len(time))
    #angular velocity
    W  = np.zeros(len(time))
    moment_ave = np.zeros(len(S1))
    indx0 = 0
    indx1 = 0

    while True:
        indx1+=1
        if indx1 == len(revolutions):
            break
        #finding the rising edge

        try:
            if revolutions[indx1] == 1:
                #determine the time to the risiing edge
                t1 = time[indx1-1]
                T = t1-t0 #period
                f = 1/T #frequency
                W[indx0:indx1] = 2*np.pi*f
                rpm[indx0:indx1] = f*60
                moment_ave[indx0:indx1] = (max(S1[indx0:indx1])
+min(S1[indx0:indx1]))/2
                t0 = t1
                indx0 = indx1
                #finding the falling edge
                while revolutions[indx1] ==1:
                    indx1+=1
                    #determine the time to the falling edge
                    if revolutions[indx1] ==0:
                        t1 = time[indx1-1]
                        T = t1-t0 #period
                        f = 1/T #frequnecy
                        W[indx0:indx1] = 2*np.pi*f
                        rpm[indx0:indx1] = f*60
                        moment_ave[indx0:indx1] = (max(S1[indx0:indx1])
+min(S1[indx0:indx1]))/2
                        t0 = t1
                        indx0 = indx1
        except IndexError:
            break

    return pedal_position, W

def Bending_Calibration():
```

```
    #calibration coefficients from Bending - take note the sign is take from the direction
    #of the calibration curve

    C11 = 54.74
    C12 = 5.05
    C21 = 5.60
    C22 = 54.28

    #the calibration matrix
    C = np.array([[C11, C12],[C21, C22]])

    L, V = np.linalg.eig(C)

    return V

def Force_Calibration(V):
    C11, C22 = 2477, 2529

    L = np.array([C11, C22])
    C = np.dot(V, np.dot(np.diag(L), np.linalg.inv(V)))
    #inverse calibartion matrix to deterime force from signal
    inv = np.linalg.inv(C)

    return inv

def angle(S1,S2):
    lamda = np.zeros(len(S1))

    i=0
    for  s2,s1 in zip(S1,S2):

        if s1>0 or s1==0:
            try:
                lamda[i]=np.arctan(s2/s1)
            except ZeroDivisionError:
                lamda[i]=0
        if s1<0:
            try:
                lamda[i]=np.arctan(s2/s1)+np.pi
            except ZeroDivisionError:
                lamda[i]=0
        i+=1
    return lamda


#sample of raw data
#contact on anthony.crimin@gmail.com for txt sample data
data =[['0', '17', '-33670.00', '-34590.00'], ['0', '20', '-33743.00',
'-34961.00'], ['0', '18', '-33884.00', '-35075.00'], ['0', '17', '-33894.00',
'-35178.00'], ['0', '18', '-33848.00', '-35693.00'], ['0', '17', '-33717.00',
'-38255.00'], ['0', '18', '-33305.00', '-40103.00'], ['0', '18', '-32711.00',
'-40368.00'], ['0', '17', '-33043.00', '-39979.00'], ['0', '18', '-33997.00',
'-39304.00'], ['0', '17', '-34162.00', '-38738.00'], ['0', '19', '-33394.00',
'-38977.00'], ['0', '17', '-33290.00', '-39644.00'], ['0', '18', '-32310.00',
'-42771.00'], ['0', '17', '-31755.00', '-44694.00']]

'''
#importing the text file with one row at a time. Save the text file were the
#python script is saved
raw = []
with open("datafile.txt", 'r') as fp:
    lines = fp.readlines()
    for line in lines:
        raw.append(line)

#removing all the delimiter - the top row is ignored using this command [1:] as
#it is the header
```

```
data = [i.strip('\n').split(',') for i in raw[1:]]
'''
#t is the delta time between each recorded data point and converent to seconds
#from seconds
t = np.array([float(i[1])*1E-3 for i in data[1:]])
# time is the absolute time in seconds and is the incremental sum of "t"
time = np.zeros(len(t))
time[0] = t[0]
for i, ii in enumerate(t[1:]):
    time[i+1] = time[i]+ii
#time in mins for graph plot
time_mins = time/60

#the Force data is the 3rd and 4th column
S1 = [float(i[2]) for i in data[1:]]
S2 = [float(i[3]) for i in data[1:]]
#a sqaure wave bwteen zero and one denates each pedal revolution - zero is the
#upstroke if the magnet is placed on the left chainstay
revolutions = [float(r[0]) for r in data[1:]]


pedal_position, W = Rising_edge(S1,S2,revolutions)
V = Bending_Calibration()
inv = Force_Calibration(V)
lamda = angle(S1,S2)

Cx,Cy = zip(*[np.dot(inv,[s2,s1]) for s1, s2 in zip(S1,S2)])

Fy = [fx*np.sin(l)-fy*np.cos(l) for fx,fy,l in zip(Cx,Cy,lamda)]
Fx = [fx*np.cos(l)+fy*np.sin(l) for fx,fy,l in zip(Cx,Cy,lamda)]

r = 0.5*208.8E-3 #m
P = [w*r*f for w,f in zip(W,Fx)]

points = 80
c1 = 0
c2 = points
P_ave = np.ones(len(P))

while c2 <= len(P):

    P_ave[c1:c2] = sum(P[c1:c2])/points
    c1 += points
    c2 += points
    #print(c2)


plt.figure()
plt.plot(time,P_ave,"--")
plt.plot(time,P)
plt.plot(time,[i*500 for i in revolutions],":")
plt.grid()
plt.xlabel("time [s]")
plt.ylabel("Power [W]")
plt.legend(["average power","instantaneous power","Rising edge - right crank at 90 degrees"])
#plt.xlim(65, 73)
#plt.ylim(0, 700)
plt.savefig("chain power")

plt.figure()
plt.plot(time,Fx,"--")
plt.plot(time,Fy)

plt.grid()
plt.xlabel("time [s]")
plt.ylabel("Force [N]")
plt.legend(["Fx","Fy"])
```

```
plt.savefig("chain force")



plt.figure()
plt.plot(time,[np.sin(l) for l in lamda])
plt.plot(time,pedal_position,"--")
#plt.xlim(77, 80)
#plt.ylim(-1.5, 1.5)
plt.xlabel("time [s]")
plt.ylabel("Sine of crank orientation")
plt.legend(["orientation","Rising edge - right crank at 90 degrees"])
plt.savefig("Sine of crank orientation")
plt.show()
```

The manufacturer's authorised representative in the EU is Springer Nature Customer Service Centre GmbH, Europaplatz 3, 69115 Heidelberg, Germany. If you have any concerns regarding our products, please contact ProductSafety@springernature.com

Printed and bound by CPI Group (UK) Ltd, Croydon, CR0 4YY
30/09/2026
02225513-0008